Kabilov S. S.

POLYMPHORIC BLOOD SYSTEMS OF THE SCHWYZ BREED

Kabilov S. S.

POLYMPHORIC BLOOD SYSTEMS OF THE SCHWYZ BREED

For the first time in the conditions of Uzbekistan blood groups of Shvitz breed animals were studied, as well as the nature of genetic processes

ScienciaScripts

Imprint
Any brand names and product names mentioned in this book are subject to trademark, brand or patent protection and are trademarks or registered trademarks of their respective holders. The use of brand names, product names, common names, trade names, product descriptions etc. even without a particular marking in this work is in no way to be construed to mean that such names may be regarded as unrestricted in respect of trademark and brand protection legislation and could thus be used by anyone.

Cover image: www.ingimage.com

This book is a translation from the original published under ISBN 978-620-8-43879-1.

Publisher:
Sciencia Scripts
is a trademark of
Dodo Books Indian Ocean Ltd. and OmniScriptum S.R.L publishing group

120 High Road, East Finchley, London, N2 9ED, United Kingdom
Str. Armeneasca 28/1, office 1, Chisinau MD-2012, Republic of Moldova, Europe
Managing Directors: Ieva Konstantinova, Victoria Ursu
info@omniscriptum.com

Printed at: see last page
ISBN: 978-620-4-18973-4

KABILOV S.S.

POLYMPHORIC BLOOD SYSTEMS OF THE SCHWYZ BREED

M O N O G R A F I C A L L Y

I. INTRODUCTION

Creation of herds with improved productivity and increased resistance to industrial conditions in a short period of time makes it necessary to take into account in breeding the data on genotype structure, metabolism and behavior of animals. First of all, the data of immunogenetics, which studied the inheritance of blood groups, are important, because this will allow to use individual features of the organism as genetic markers in animal breeding.

The aim of the study is to establish the immunogenetic features of Schwycki cattle bred in Uzbekistan and the possibility of using them in breeding.

In accordance with this goal the following tasks are set:

1. To study the genetic polymorphism of erythrocyte antigens, to establish the alleles controlling them and the possibilities of using them as genetic markers.
2. To study the relationship between blood groups and milk productivity (milk yield, milk fat content) of animals.
3. To determine the possibilities of using blood groups to increase the fertility of cows from the first insemination.
4. To confirm the effectiveness of immunogenetic control of the reliability of the origin of breeding animals.

For the first time in the conditions of Uzbekistan, blood groups of animals of the Shvitz breed, as well as the nature of genetic processes in the population of Shvitz cattle in the process of selection have been studied.

The dependence of the level of milk productivity of cows-daughters on their inheritance of alternative paternal alleles of blood groups has been established. The influence of homo- or heterozygosity of cows by blood group systems on the degree of their fertility was revealed.

The study of genotype features of Shvitz animals will allow tracing the dynamics of blood group allele frequencies as a result of acclimatization. This is important because the current practice of importing animals of brown breeds to Uzbekistan from other regions of the country becomes economically unprofitable. Imported animals take a long time to adapt in extreme conditions of the republic and quickly lose productivity, especially in hot season. Hence the relevance of improving local, long-bred and creating new more highly productive, suitable for use in industrial technology lines and types of animals.

Organization of selection of shvitzized cattle on a scientific basis will allow to identify the most valuable related groups of animals, transform them into lines and thus accelerate the process of breed formation of brown cattle of Central Asia. Selection of mating individuals with consideration of genotype peculiarities will allow to increase heterosis effect purposefully and to develop proposals for predicting milk productivity of animals.

In herds of the three farms under study (breeding farms "Savay, Izbaskan and "Gallyaaral"), periodic examination of the origin of breeding animals based on data on their blood groups is carried out.

Evaluation of breeding bulls is based on the productivity of true daughters with proven blood group ancestry.

The selection of breeding stock with alleles of blood groups positively related to milk productivity indicators is carried out.

All these measures allow to complete the herds of breeding farms with highly productive animals, accelerate the creation of dairy shvitzized type of cattle of Uzbekistan.

1.1 The Schwyz cattle breed: distribution and productive features

Schwycki cattle belongs to the group of brown breeds and is widely spread not only in Europe, but also in the whole world. According to F. Haring and H. Messerschmidt, there are more than 170 thousand cattle of the Schwyckian breed in the USA, which is 6% of the total livestock.

Unlike the Swiss, the Schwyz cattle of the USA are closer to the dairy type, larger and physically stronger than other breeds. At the age of 4-5 years, the live weight of cows reaches 670 kg, and bulls - 900 kg. Animals of this breed are characterized by good health, high fertility and adaptability to different climatic zones

According to F.Haring, V.Pavlov /117/, A.S.Vsevolkikh /35/, Schwyck breed cows surpass Ayrshire, Jersey and Red Danish breeds in milk yield. In some herds, the average milk yield is 6995 kg with milk fat content of 4.2%.

Animals of this breed are characterized by longevity. So, a significant part of cows in the USA was used during 10 lactations with average milk yield of 9753 kg and milk fat content of 4,64%, i.e. they gave 488 kg of milk fat.Shvitsizirovanie cattle gave their results. Thus, in the breeding farm "Karavaevo" in 1951 the average milk yield from a cow amounted to 6406 kg, with the presence in milk of 3,73% fat. Bulls of the Schwitz breed were used in the creation of the domestic Kostroma breed, which was approved in 1944.

In eight provinces of Uzbekistan Shvitz cattle is a planned (improving) breed. In the leading farms of the republic these animals have good productivity in the range of 2700-3200 kg of milk with fat content of 3.7-3.9%.

In the republic by crossing local zebu cattle with Schwyz breeding bulls and derivatives of these breeds a large array of highly productive Schwyzized animals, well adapted to the conditions of sharply continental climate of Central Asia and industrial technology, has been created to date.

1.2 Bovine blood groups: breed and lineage differences.

In 1900, the Austrian physician K. Landsteier established individual differences in human blood, thus initiating the study of molecular forms of variability. A few years later, blood groups were also found in various breeds of farm animals. For the first time data on the presence of blood groups in cattle and the peculiarities of their distribution to breeds was obtained by S. Ferguson. It was confirmed in further studies. According to 30 antigenic factors known at that time, differences between animals of the Gernseys and Holstein-Friesian breeds were revealed.

From the data of A. Neimami-Sorensen, who studied blood groups of three Danish breeds, it is known that 75% of blood group alleles up to B-system found in animals of the Red Danish breed are absent in Jersey breed, and 85% of alleles of the same system of Jersey breed are absent in Burns Red Danish breed. In cows of the Swedish breeds of the Red-Friesian, Friesian blood group in 1200 animals, 23 alleles of the B-system were detected in cows of the Red-Friesian breed, 8 alleles in animals of the Friesian breed. On the same livestock were found and significant interlinear differences studying in a comparative aspect of blood groups of Simmental cattle with American and European breeds, a significant difference in the B-system. Thus, half of 40 alleles of this system were not found in other breeds.

Some alleles of the B-system, absent in other breeds, were detected in red Polish cattle whose blood groups were studied by A.Spryszak, which allowed the author to draw a conclusion about the genetic individuality of this breed. At the same time, the author determined that Polish and Danish Red cattle are related breeds based on the distribution of blood groups.

The genetic structure of two Dutch breeds - black and red-breed was studied by Bouw, who found that despite their similarity in a number of exterior and productive traits, there is a significant genetic heterogeneity between them, hereditarily determined by blood group loci.

Specificity of brown cattle by blood groups was discovered by N.Buschman when studying a number of German cattle breeds (mountain spotted, brown, yellow, Pinzgauz, etc.).

In the German black-and-brown breed O.Richter revealed 120 phenotypes according to systems B and 0, while in brown cattle there were 75 of them, and in pale cattle - only 30.

According to K.Maiyala, S.Lindstrom, the greatest difference between Ayrshire cattle and Finnish cattle was revealed by the frequency of antigenic factors of the systems: P, U, I', J2. Thus, they came to the conclusion that in Ayrshire cattle alleles B^{BG0}, $B^{Y}2^{E}3^{G'}$ are more frequent , B in $°1^{A'}$, $B^{in}3$ $B^{Y}2^{(E)}1$, and ^{in} Finnish $B^{A}1$, B^{B}, $B^{(BGKE)}2^{°'}$, $B^{G0}1^{(Y)}2$.

When analyzing blood groups of high-mountainous breeds of animals in Germany, D.O.Schmid found that yellow cattle have 30 alleles of B-system, brown cattle have 68 alleles, high-mountainous spotted cattle have 113 alleles and Pintsgauz cattle have only 9 alleles.

The study of blood group polymorphism can reveal not only differences, but also similarities of animals. In this way, for example, D.R.Osterhoff established kinship between South African and some European breeds.

In the study of 5 American dairy breeds on the B system of blood groups K.Stormont /140/ found that in Ayrshire cattle the greatest frequency had alleles B $°1^{A}1$ (0.22), $B^{B0}1^{Y}2^{D'}$ (0.14), B °1 (0.11) B $°1^{(Y)}1^{E}3^{G'(Y)'}$ (0.10),B $°3^{(qy)(')k')('(o)(}$0.14); Holstein-Friesian cattle-B $^{BY}2^{E}1$(0.12), B $°3^{QJ}2^{K'o'}$ (0.06), B (0.06);Gernsey cattle-B $^{10}x^{E}1$ (0.23), $BBGKO_{X}E_{F}$'O'(0.(13); in Jerean cattle, B BGKOXY1A1E3G'K'B' O'$Y^{(1)}$B°

1T1Y1E3x$^{(E)}$3$^{B\circ\ 1}$ 0.14); in Schwyz cattle, B °1^{T}1$^{(Y)}$2$^{(E)}$F (0.37) , B B0$_{(2}$$^{)Y}2^{A,E(,)(G)(,)(,)(Y)'}$ (0.10).

When analyzing the blood groups of Swiss cattle, J .Reuse identified 65 B-system alleles. The most frequent alleles were BPQE2'BBP,B GY$_{2E}$A,D,B especially in Eringa cattle. The degree of homozygosity of animals of this breed was only 0.05, which indicates a significant level of its genetic variability in the population.

According to J.Oldenbrock, J.Bouw, as a result of active use of sons of bulls with genotype $^{B\ (B0)}$ the frequency of the latter has significantly increased in animals of the Dutch red-bred breed.Significant genetic diversity in the B-system was found in animals of the Hungarian mottled breed (NO alleles). The most frcquent alleles were B ^{G}1^{A} (0.1077), c°1 (0.0720), ^{BY}2$^{A,(D),B'}$(0.0620), c5°1^{T}1$^{(B)}$3$^{P'}$ (0.0452) and B°2 (0.0414$^{)}$.

The diversity of blood groups in animals of Lithuanian black-fur, Lithuanian red and Latvian brown breeds was studied by 3.Vagonis and his colleagues /19, 20, 21, 22, 23/. The first of them had significant differences from the others by this characteristic. Lithuanian black-mottled cattle of Lithuanian origin have increased frequency of alleles in x , in , in , in , in 2.

The peculiarity of blood antigenic factors distribution in Estonian black and red breeds is the absence of I 1 and J1 antigens in the former. In red Estonian cattle 102 alleles were detected in the B system, while in black-breed cattle there were only 67 alleles /L.F.Vaher, 25, 26/.

Data on the immunogenetic features of the Holmogorsky, Black-Beast, Sachevsky, Shvitsky, Caucasian Brown and Alatau breeds in Russian breeding farms were obtained by P.F. Sorokovoy et al. /133, 134, 135, 136, 137, 138/. The complete absence of antigenic factor M in animals of Shvitz, Caucasian brown and Alatau breeds was found. In

Holmogor cattle it was found in 27% of animals, and in black-and-brown cattle in 18%. The distribution of B-system alleles was as follows: in black-and-brown cattle there were 105, in Kholmogor cattle - 80, in Shvitz cattle - 75, in Caucasian brown cattle - 52, in Sychevsk cattle - 94 and in Alatau cattle - 79. On the basis of the found common 35 alleles of the B-system it was concluded that the Shvitz, Caucasian brown and Alatau breeds are significantly related.

In the red Gorbatovskaya breed 82 alleles were identified by the B system, of which 26 were frequently occurring alleles with a total frequency of 0.8604 /Sorokovoy P.V., Slepchenko A.P. 135/.

Immunogenetic characterization of animals can also be used in the study of genetic peculiarities of pedigrees of different farms of one region. Thus, studies by S.Shadmanov /146,147,148, V.P.Pavlichenko 116/ and others, carried out in the Leningrad region and Karelia, established that in the herd of the breeding farm "Petrovsky" the frequency of occurrence of allele 0g is higher than in the herd of other farms, and antigenic factors z, Q, y are absent in the animals of the breeding farm "Lesnoe". The alleles G1O3P1T1Y2 , P1J1J1J1j1jS1S2 occur in the population of the breeding farm "Lesnoe" with a higher frequency than in animals of other herds. In addition, in the corresponding analysis of the B and C systems, 31 alleles of the B-system out of 52 were common in the herd of black-breed cattle for the mentioned two breeding farms.

The studies of N.F.Anisimov /5/ indicate that Simmental and Sychev breed of cattle have a significant kinship. This conclusion is also confirmed by the data on blood groups.

The diversity of blood group genotypes of Simmental cattle in Ukraine was the subject of studies by V.Y. Meshcheryakov /104/ and M. Mashurov /94/. Thus, in the study of Simmental cattle in the breeding farm "Elansky" in Voronezh region and the Gosplemzavod named after Lenin in Tambov region. A.M.Mashurov established that 12 alleles of B system were common for herds of both farms, the index of genetic similarity amounted to 0.68

Comparison of population features of Sychiv and Simmental cattle was carried out in numerous studies by P.F. Sorokovoy and coworkers /136, 138/. The diversity of blood groups according to the B system in Sychev cattle was represented by 77 alleles, and in Simmental cattle population - by 74 alleles.

The difference between these breeds was reflected in the absence of $BY2A2E1^{D}$ alleles in Sychiv cattle. Differences were also found in rare alleles, e.g. alleles $^{B(B)}1^{T}1^{TA}1B^{(1)}D$, $_{B}B_{2}{}^{G}3^{T}1^{A}1^{B,E}3^{p,Q}$, BBIE3G alleles were detected only in Sychev cattle, and alleles $B^{(c)}1^{\circ}1^{t}1^{(E)}Z^{f,k}$,$B^{(G)}3^{G}1^{(T)}1^{E}3^{(G),K,(G)}$, B01. B only in Simmental breed herds. The index of genetic similarity calculated by the formula of L.A.Zhivotovsky et al. /59/, amounted to 0.779.

The Yaroslavl breed of cattle of the breeding farm "Gorshikha" was studied by V.F.Maksimenko /88/ during three generations of selection, having established a significant difference in the concentration of a number of B alleles.

K.Waiawski studied the black and mottled cattle breed of the Polish People's Republic and found that the change of blood group alleles of animals occurs as a result of artificial selection. It was noted that individuals with alleles $B^{Bo}1$, $BB^{0}xAP$ and some others are subject to elimination. In addition, he found a decrease in the total number of B alleles from 176 to 118, which J.Treia explains by the reduction in the number of bulls-producers.

These results were confirmed in the studies of A.Kaczmarek and Z.Dorynek /72/, who studied blood groups in black-breeding bulls in Poland. During 16 years of observations, the frequency of alleles $B^{G}2^{Y}2^{E}1Q$ and $B^{c0}1$ in the group of bulls left in reproduction increased sharply, while the frequencies of BG2KOxY2A'O'Z and BY 2'D'G'Z'Q decreased significantly.

The opposite results were obtained by D. Kovacs and E. Takacs during the 11-year period of observations of genetic structures of blood group loci. They revealed an insignificant change in the frequencies of these alleles in Hungarian gray cattle. The difference in occurrence was unreliable and did not exceed 4%, although the herd was updated by 75%. The value of the degree of homozygosity was almost unchanged.

Data on changes in gene frequencies over a 10-year period of research of Kholmogorsk cattle of the breeding farm "Lesniye Polyany" are given by E.G. Vorobiev and P.F. Sorokov /32/. In 1968, 66 alleles of the B-system were identified in the herd of the farm, and in 1975 - 45 alleles, which indicates a significant decrease in the level of variability over two generations. Out of the number of alleles identified, only 12 were the most widespread. The total frequency of occurrence of the latter amounted to 0.6368, while in 1975 it was 0.7031. At comparative comparison of genealogical lines and kin groups of the breeding farms "Lesniye Polyany" of Moscow and "Kholmogorsky" of Arkhangelsk regions, no significant differences between them were noted. In groups of bulls-producers of breeding farms of Arkhangelsk and Moscow regions the highest concentration had 10 alleles of B-system with a total frequency of 0,7373. These data, obtained in a study of 278 head, are cited as evidence of higher homozygosity of bulls compared to cows.

It follows from the above data that studies of bovine blood group polymorphism have proved that it is possible to differentiate lines, families and breeds using markers. Each breed has a specific set of blood group loci, but by the presence of common loci it is possible to establish the degree of their kinship. The variation of blood group allele frequencies is used to study selection and genetic processes in farm animal populations.

1.3. Immunogenetic characterization of the Schwyck breed of cattle

The Schwyz cattle belongs to the group of brown breeds. As already mentioned in section 1.1, it has a very wide distribution all over the world and is widely used in breeding other breeds. In particular, the Shvitz cattle was used in the development of the Kostroma breed, and the Kostroma breed in turn participated in the development of the Alatau and Caucasian brown breeds.

In the world literature the question of hereditary features of the Schwyz cattle breed by blood group loci is reflected quite completely.

Thus, the blood groups of the Schwyck breed of cattle in the leading state breeding farms "Pakhomovo" of Tula and "Aleksandrovo" of Moscow region were studied by E.G. Vorobyov /31/. In the herd of the first farm he identified 50 alleles in the B system, and in the second farm only 30 alleles. The genetic similarity index was only 0.29, which indicates a significant genetic difference between them.

A.K.Borozdina et al. found 64 B-system alleles in the herd of the Aleksandrovo breeding farm a few years later. /15/ found 64 alleles of the B-system, 18 of them belonged to the frequently occurring ones, their total frequency was 0.6801.In the herd of Schwyz cattle of the Kominterna breeding farm P.P.Kochetkova /75/ found 61 alleles.

The alleles of blood groups, which are characteristic only for the Shvitz breed, have been revealed; when comparing the populations of Shvitz and Kostroma cattle, common alleles have been found in them, confirming the genetic similarity of these breeds.

In the works of P.G. Klabukov et al. /73/ compared immunogenetic features of cattle of brown breeds. In particular, it was revealed that B alleles° 1T1E3F$^{K'}$ are widespread in Shvitz cattle, but they are almost absent in animals of the Kostroma breed. Allele B is rare in the Shvitz breed. According to the data of A.S. Vsekikh /38/, allele B $Q^{A}1^{E}1$ is also rare in the Shvitz breed.

Immunogenetic markers can be used to detect not only interbreed but also interlinear differences. Thus, studies by G.M.Aleksandrova, T.V.Bakhmutova /30/, A.S.Vsevolkikh et al. /42/ found that in the breeding farm "Pakhomovo" in Shvitz cattle allele B was often found in the line of the bull producer Boyarin, allele B $^{\circ}1^{T}1^{Y}2^{E}h^{F}2$ in the lines Almaz and Baro, allele BO1T1E3F'K in the lines of Dick and Zikola, B I1Y1E 111 O Y D'G' in the lines of Boyarin and Mirny, B 1 2-Y'G'Enkel and Baro, allele B ^{1} in the lines of Almaz and Dick.

Thus, the study of blood groups of the Shvitz cattle revealed its immunogenetic individuality not only to the breed as a whole, but also in smaller individual structural units.

1.4.Immunogenetic basis of cattle breeding

Reliability of origin and well-established records are the basis of highly effective breeding and pedigree work. The presence of errors in breeding records can dramatically reduce the impact of many years of labor-intensive work to improve existing and create new breeds, lines and hybrids.

More than twenty years ago, biochemical (serum proteins of blood and milk) and immunogenetic (polyvalent sera, monospecific antiserum) markers of individual variability began to be used in addition to traditional methods of animal origin recording.

The most effective, according to C.Stormont , is a method that combines biochemical and immunogenetic markers.

The new method of controlling the origin of animals is based on the principle that the blood groups found in the offspring must be those of its parents and passed on to the offspring in accordance with

Mendelian laws. The validity of the record is established by the antigens of the parents and offspring (phenotype) and by the alleles that make up the genotype (genotype). Alleles and genotypes are established by family genetic analysis by comparing the blood groups of parents and offspring. For some immunogenetic systems, the genotype of animals can be established directly from the data of serologic tests. Such systems include F and R. Antigens are transmitted as combinations in such complex systems as B, C and S. The genotypes established are then used to analyze the validity of the animal's provenance record.

A normal individual has two alternative alleles at each locus, one of which she gives to her offspring. If an offspring is found to have an allele missing in the parents, the paternity or maternity recorded in the pedigree book is excluded. The most effective is regular control of the reliability of the origin of animals of each generation, i.e. when it is possible to trace the process of allele transmission from father to son, grandson, etc.

According to Z.I.Vagonis and co-workers /19, 21/, the efficiency of controlling the origin of breeding animals depends on the number of reagents in the antiserum set used in the tests, especially for B and C-systems. In addition, the reliability of the assay depends on the heterogeneity of populations, breeding methods and the quality of antiserum obtained or purchased.

These authors believe that with the help of only 40 reagents for 10 immunogenetic systems of blood groups it is possible to establish accurate data in 96% of cases of disputable origin of breeding animals.

In domestic and foreign literature there are many works concerning methods and efficiency of control of the origin of breeding animals /Bagrii B.A., Meshcheryakov V.Y. /7/; Vagonis Z.I. /19, 21/ etc.; Vyaskikh A.S. et al. /38/; Kushner H.F. /81/; Lapushkova V.V.

/84/; Semyonov V.A., Sorokov P.F. /129/; Slepchenko AP. 35/; Tikhonov V.N. /147/; Shadmanov S.I. /145/; Ykva T.Y. /154/; Eidrigevich E.V., Lapkin M.M. /155/.

According to A.M.Mashurov's data, origin control in the Lenin and Elansky breeding farms was carried out on 2454 breeding bulls and heifers. The percentage of errors in the offspring of individual bulls ranged from 0 to 63.7%.In the Smolensk region, during the study of blood groups of 828 animals of the Sychev breed, N.F.Anisimov /4/ revealed from 21.1% to 31.0% of offspring with unreliable records of paternity and only 1.0 - 1.8% - with unreliable maternity.

P.G.Klabukov and P.F.Sorokovoy /73/ cite data obtained from breeding farms, where 0.25-30% of animals had an unreliable record of origin.

According to V.A.Semenov and P.F.Sorokovoy, in 1976 in the breeding farm "Proletarian" unreliable paternity was detected in 20.9% and unreliable maternity in 8.9% of 360 investigated progenies. In 1974, in the Karavaevo breeding farm, in a study of 108 cows, inconsistency of the recorded father was found in 18.4% and mother - in 4.3% of animals.

According to E.P.Vagonis, the reliability of origin of 126 breeding steers in the Lithuanian SSR was only 12% during their inspection. In the study of 407 young bulls from four farms of the republic, on average 42.3% of the progeny did not receive confirmation of the authenticity of origin.

Among the most common causes of erroneous pedigree entries, Koch lists the following: loss of tag or overgrowth of ear number; participation of two or more bulls in the insemination of one cow; substitution of calves at birth; re-covering of the uterus by a bull; incorrect entry in the registration book, replacement of semen ampoules.

A.Tolle K.Beuche considers the causes of false records in more detail and subdivides them into two groups: errors due to negligence of maintenance personnel and errors based on biological features of the breeding process.

According to J.Rendei, in Sweden, out of 5000 cows that were artificially inseminated, only 7.3% were re-inseminated with semen from the same bull. But in this case, it should be taken into account that the period of pregnancy itself and the duration of the hunting cycle vary widely in different cows.

According to P.Mitscheriick, L.Tolle the intervals between two adjacent hunts have the greatest influence on re-insemination. Thus, if the interval is 21 days, up to 6% of calves can come from both the first and the second bull. If the interval between two hunts is 18 days, the number of such calves increases to 11%, and if the interval is 15 or 12 days, these figures reach 18% and 28.5%, respectively.

According to A. Vinikas /27/, even farms *with* well-prepared breeding records are not immune to errors. The most probable reasons for them, according to the author, are as follows: accidental confusion of semen in the farm (up to 34%) and at the breeding enterprise (up to 4%); repeated insemination with semen of different bulls-producers of one cow during one hunt (up to 6%); fertilization of an already inseminated cow that came in a false hunt (up to 8%); confusion of calves during tagging or breeding of several cows (up to 6%); miscalculations in keeping breeding records (up to 6%).

When evaluating bulls by progeny quality, a high percentage of unreliable progeny, especially daughters, negatively affects its reliability. Thus, according to P.F. Sorokovoy and A.R. Slepchenko /135/, there were significant discrepancies when evaluating bulls by reliable and nominal daughters of the Zimenki breeding farm and the Voykov breeding farm of the Vladimir region.

Similar results are reported by P.G.Klabukov et al. /74/.In their research, when analyzing 15 daughters recorded for the bull Dar 5206 in the breeding farm "Dugino", the average productivity for the first lactation was 2848 kg of milk at 3.55% fat, while his seven true daughters had a milk yield of 3270 and a milk fat content of 3.66%.

A.S.Vyaskikh, who conducted similar researches in the breeding farm "Proletarian", also points out the necessity to evaluate bulls only to his reliable daughters /40/.

A significant number of works in domestic and foreign literature is devoted to the study of the relationship of blood groups with economically useful traits of farm animals (Bernikova N.N., Shadmanov S.I. /12/; Vagonis Z.I. et al. /20/; Vaher L.F. /25/; Gontov M.E. /50/; Malkevich V.M. /89/; Okhapkin S.K., Khrunova A.I. /27; Sorokovoy P.F. D35/; Shadmanov S.I. et al. /145/.

At present, there are a number of hypotheses regarding the mechanisms of association of blood group loci with productive traits. Thus, I.Johanoen, I.Rendel, O.Gravert/67/, Mashurov A.L./93/, L.S.Zhebrovsky, V.Litjutko /58/, Shadmanov S.I./146/, E.V.Eidrigevich /157/ propose the following genetic mechanisms of association between blood group loci and economically useful traits of animals: I) pleiotropy; 2) genetic coupling of loci; 3) heterosis effect.

Pleiotropy is a type of interaction between genes, when one of them can cause two or more economically useful traits. The cohesion of genes is manifested in the fact that they, having joined together in the process of evolution or selection into a kind of block, are then inherited, and with them the traits defined by them

Some authors point out the negative effect of allele I on milk fat content - J.Rendel , L.F.Wacher et al. /25/, others on milk yield (W.H.Rausch , L.F.Wacher and others /25/).

According to W.Weber, who conducted studies on Danish black-breed cattle, the presence of allele B^E3 is accompanied by an increase in productivity per lactation by 400 kg, while according to P.G.Klabukov and P.F.Sorokov /73/ a negative correlation between its presence and milk yield was found. S.I.Shadmanov et al. /146/ report that carriers of this allele were superior in milk fat content of animals without the B E3 allele.

The positive influence of allele $B^{B0}1^{YD'}$ on the level of milk productivity is substantiated in the works of A.Neimarm, Sorensen, A.Robertson, J.Rendel, R.M.Coneally, W.H.Stonewho conducted a study on animals of different breeds.

In a herd of Holstein-Friesian breed N.Rausch, B.W.Brum, T.M.Ludwick revealed a positive association of B allele BO1Y1D with milk yield. Cows with this allele were 305 kg higher in milk yield than animals without it.

L.F.Vaher et al. /25/ in the study of the red Estonian breed, on the contrary, found no favorable effect of the allele $B^{BO}1^{Y}1^{D'}$ NEITHER on milk yield nor on milk fatness. According to their data, cows carrying this allele had milk yields 354 kg lower than animals lacking it.

According to V.J. Dexne et al. /57/, cows of brown Latvian breed had a difference in productivity depending on the heterosis effect is manifested in the predominance of offspring over parents due to the fact that their genotype contains more genes in heterozygous state.

J.S.Ferguson pointed out the connection of blood group loci with the level of milk productivity. The same opinion is held by M.R.Jrwin. Hendel denies complete neutrality of genes controlling blood groups in relation to productive traits. N.V.Tikhonov considers the relationship between blood groups and productive traits as "pseudo-correlation".

Sometimes the progeny of individual improving bulls have alleles in their genotype that are not found in the progeny of other bulls, and these alleles do not necessarily represent markers of high productivity.

Such correlations E.V.Eidrigevich et al. /157/ called "false linkage". As a result of long-term intensive use of such bulls, saturation of their progeny with alleles of certain combinations occurs, which eventually leads to the division of the herd into heterogeneous micropopulations.

The study of milk productivity of cows in connection with the difference of their genotypes on loci of blood groups in some cases gives contradictory data. Thus, analyzing the level of milk productivity in connection with the presence of the M allele in black and sow cattle, some authors - E.Jiitecherllck, A.Tolle, E.Walter, I.N.Bernikova /11/ - stated that carriers of this allele had a lower miracle than those animals that did not have it. In the studies of other authors on Kholmogorsk cattle, for example, P.F.Sorokova /136/, it is indicated that cows with the M allele exceeded the milk yield of animals that did not have it. In the studies of P.G.Klabukov /73/ no differences in milk productivity in cows were found due to the presence or absence of their genotype by blood groups. Thus, animals with allele B $°2^{A}2^{U'K}$ IN milk yield for the third lactation were significantly superior to animals that did not have it, and carriers of allele B vr showed superiority in milk fat content.

According to N.N.Bernikova and S.I.Shadmanov/12/, a certain tendency to increase or decrease the level of productivity in animals with different alleles was maintained during several generations of selection. In the study of blood groups of one of the ancestors of the Dutch black-and-brown breed, M.Zurkowski, J.Bouw pointed out that milk productivity of offspring is in direct dependence on the alternative paternal alleles inherited by them. It was found that daughters who received the B $^{1}2$ allele at locus B of the blood group system were characterized by a higher percentage of fat in milk than carriers of the alternative allele B1. One of the ancestors of the Schwyz breed, the bull Almaz, according to P. G. Klabukov and P. G. Klabukov.G.Klabukov and P.F.Sorokov /73/, those daughters who received the paternal allele B $^{\circ}1^{T}1^{Y}2^{E}3^{F}2$, exceeded the milk yield of their half-sisters with the alternative paternal allele B $^{c}3$ in the first lactation by 394 kg and 343 kg in the second lactation, respectively.

The work of A.R.Slepchenko provides data that the daughters of the bull Bodrogo, which received allele B, exceeded their half-sisters with allele B 2 1 to milk yield by 677 kg and milk fat content by 0.2%. In the same herd, daughters of the bull Roma 2702 with allele BY 2Y had 0.82% higher milk fat content than offspring with allele B $^{\circ}1^{t}1^{Y}2^{c}h^{F}2G$ (P<0.01).

Hereditary factors of the father, as L.K. Ernst, P.F. Sorokovoy and A. Mashurov believe, transmitted in one or another chromosome, can sometimes have different effects on milk productivity. Thus, in a herd of black-and-brown animals, the daughters of the bull who received marker allele B1 from the father's chromosome were significantly superior in milk yield to their half-sisters who inherited the chromosome with marker allele BDY 2GPG. Daughters of bull 353 with marker B2 had an advantage in milk fatness over their half-sisters with marker $c^{BG(K)(Y)}2^{\circ}$.

In studies on Kostroma cattle, A.S. Vyaskikh /36/ noted that daughters of heterozygous bulls inherited from their father the factors represented by the alternative B allele were significantly distinguished by the level of milk productivity. Thus, the daughters of the bull Mrak from the line Karo with allele B 1 surpassed his daughters with allele B in milk yield per lactation by more than 450 kg (B < 0.05). In the same herd, milk yield of daughters of Salyut with Bp1E3I allele was 602 kg higher than that of daughters with marker allele BB2QT1FIP.

In the study of blood groups of Alatau cattle Bykovchenko B.G. et al. /18/ revealed that daughters of bull Novy 533 with allele G2 E3'F'O allele B 2 3 had milk yield in the first lactation by 371 kg lower than daughters with allele B $^{\circ}1^{T}1^{E}3^{E\ k}$, and in the second lactation - by

520 kg and in the third lactation - by 1407 kg. The milk yield of the daughter of the bull Eretz 1419 with allele B 2 in the first lactation was 223 kg lower than that of daughters with allele B, and in the second lactation - by 246 kg. At the same time, the authors point out that daughters of other bulls, also inheriting alternative alleles of the father, did not differ significantly in productivity from the average index for the lines of the herd of animals of the Sychiv breed, as V.K.Lernushenko et al. report. /144/, the daughters of the bull Liker 586, who received from their father the allele $B^{G}3O1^{t}1^{\prime A,}2^{c}3^{F'I'K}$, had a reliable advantage in milk yield per lactation over half-sisters with the allele $B\ ^{G}1^{\circ}t$; it was also noted that the daughters of the bull Arafon, who inherited the same allele, retained this superiority.

E.G.Vorobyev noted /31/ that the difference of genotypes affects the live weight of cattle. As the results of his research showed, the difference of parents in the amount of certain hereditary factors of blood groups was reflected in calf growth.

Daughters from parents with different B-alleles were slightly superior in live weight to heifers whose parents had common B-system blood group alleles. Special attention was paid to the influence of homo- or heterozygosity of blood group loci. For example, homozygosity for the locus led to the fact that the live weight of these animals at the age of 6-18 months was 17-19 kg lower than that of peers heterozygous for the F allele. The difference was significant at 6 months of age.

S.I.Shadmanov and N.N.Bernikova studied the relationship of blood groups with the intensity of development of young animals from birth to 18 months of age in the herd of "Lesnoe" breeding farm in the Leningrad region. It was noted that carriers of the BBO3Y2E3P allele had a lower live weight at birth than calves lacking this allele. This difference persisted until 18 months of age.

All productive traits of farm animals are inherited according to the principle of polymery, and therefore, studies in which the cumulative or additive effect of several loci on them was studied are of special interest. Among the works devoted to the cumulative effect of several markers on productivity traits, it is possible to note the studies of E.G.Vorobyov /31/, A.Mashurov /94/ N.G.Bukarov /16/, V.L-Dexne /57/. The common disadvantage of all these works is that they are based on studies of one generation of animals.

An attempt to study the nature of genetic correlations was made in a series of works by A.A.Istomin et al. /70/. According to these authors, the diversity of manifested traits is in direct dependence on the number of additive and non-additive genes.

When studying this question, one should not forget that the additive effect will be manifested only if the heritability of the trait being selected is high (R.M.Conneally, Stone)Long-term line breeding leads to the fact that a certain set of additive genes will be concentrated in the genotypes of animals of different lines. In particular, A.L.Mashurov /93/ explains by this tendency the fact that positive traits found in one population are sometimes absent in other populations.Since fertility and viability traits have a low inheritability coefficient, a significant heterosis effect can be expected when mating individuals differing in blood group genes. Thus, 0.J.Stormont, R.D.Owen proved the existence of dependence of viability of young animals on the degree of similarity of mating parents by blood groups. Studies by G.I.Ivanov /60/ show that more outbreeding is observed in cows mated with bulls having similar antigens. Taking into account the indices of genetic similarity (heterozygosity), R.M.Conneally found that when the index of heterozygosity increases from 6 to 45%, embryonic fetal mortality is expected to increase from 6 to 31%.

In the last decade and a half, many interesting works have appeared devoted to the study of similarity indices between outcomes and individual herds (Mashchurov A.M. /95/, Shadmanov S.I., et al. /150/, Dexne V.Y. /57/, Ohapkin S.K., et al. /11O/, Goncharenko G.M. /49/, Deeva V.S. /56/, Baranov A.V. /8/.

When studying the reproduction function of the herd depending on the level of heterozygosity D.R.Osterhoff found a reliable relationship between it and the optimal service period.The above material shows that with the help of blood groups it is possible to give a genetic characterization of breeds and fix the processes occurring in animal populations.Some loci of blood groups can serve as markers of productivity.

At the same time, it should be noted that these issues are insufficiently studied in terms of widespread application of genetic methods in breeding and breeding of farm animals.

II.MATERIAL AND METHODS OF RESEARCH

The object of the study were maternal and daughter herds of the breeding farm "Sawai" of Kurghantepinsky and the breeding farm "Izbaskan" of Izbaskan districts of Andijan oblast and the herd of the breeding farm "Gallyaaral" of Gallyaaral district of Jizzak oblast. Blood groups were determined in bulls-producers of Andijan regional state breeding association, as well as in a group of imported animals, the seed of which was delivered from the sperm bank of CSIO. In addition, immunogenetic features of blood groups were studied in local Schwyz cattle.

In total, blood groups were determined in 50 bulls and 2115 cows and young cattle.

The modern herd of the listed breeding farms mainly consists of animals of new lines and related groups of desirable type of local selection (64-75%).

In their majority, these herds consist of purebred animals, as well as animals of the 1st generation. The farms, where the research was conducted, are the leading ones in Uzbekistan on the use of schvitzing. The average milk yield of animals here is 2600-3,800 kg with fat content in milk 3.7%-3.9%, their live weight reaches 470-500 kg. Herds of these farms have a positive impact on the whole array of schvitzized cattle in the region.

A certain part of the cattle population in Uzbekistan is made up of descendants of American bulls of the Shvitz breed, imported from the RSFSR *and* constituting breeding material.

2.1.History of farm establishment and peculiarities of herd formation

Plemkhoz "Savay" was organized in 1930 on the lands of state land reserve of Uzbek and Kyrgyz SSR. The total land area is 9310 ha, including irrigated - 8900 ha. Besides, the breeding farm uses long-term distant pastures in the tracts "Bet Munak" and "Surtak" on the territory of Kyrgyz SSR. The total area of pastures suitable for use is 4190 ha.

Average precipitation in the area of the breeding farm location is 323-4430 mm with maximum in winter-spring period (87%). The maximum air temperature in summer time is 41°C. Soil cover is represented by typical gray soils.

Plemkhoz "Izbaskan" was organized in 1932, located in the north-eastern part of Izbaskan district, 16 km from the regional center (Andijan). The main direction is cattle breeding. Big Fergana canal named after U. Yusupov and Pakhtaabad canal pass through the territory of the farm. 16 thousand people live in the farm, 24.6% of cotton, 11% of cocoons, 34% of meat, 26.9% of milk and 12% of eggs produced in the district.

The territory of the "Gallyaeral" breeding farm is 39991 ha, including 11289 ha of pastures and 21.1 thousand ha of crops. The climate is sharply continental, the summer is very hot, the average air temperature in this period reaches 36.5°C, with fluctuations from 14°C -27°C at night to 34-40°C in the daytime. This leads to burnout of natural grass cover by the end of May - beginning of June, when only late-maturing vegetation continues to vegetate, and grain crops (wheat, barley) ripen well. Yield of pastures is 3-3.5 c/ha of dry fodder mass, in some years it reaches 6.7 c/ha.

The milk yield for the last lactation in the breeding farm "Savai" averaged 3273125 kg for first heifers, 3321135 kg for second calving cows, 3324134 kg for third and older calving cows. In the herd of "Gallyaaral" breeding farm the same indicators were 2368+38; 2622±25 and 2804+-18 kg, in the herd of "Izbaskan" breeding farm - 3248±24; 3350+-36 and 3380+-31 kg, respectively.

These data exceeded the breed standard in first calving cows by 418-1323 kg, in second calving cows by 222-950 kg, and in third and subsequent calving cows by 104-680 kg. The milk yield of full-aged cows of the breeding nucleus in the "Savai" and "Izbaskan" breeding farms was 124-48 kg higher than the standard, and in the "Gallyaaral" breeding farm it was close to the standard.

The most productive animals with milk yields of 4109+-68 and 4938+-69 kg are concentrated in the group of bull-producing cows of the studied farms. In "Savay" breeding farm, cow Marta 6984 produced 5233 kg of milk with fat content of 3.7%, Venus 7906 produced 5003 kg of milk with fat content of 3.7% and Chaika 1404 produced 4421 kg. In the herd of the breeding farm "Gyallyaaral" from the cow Smela 2046 5222 kg of milk with fat content of 3.7% were obtained for lactation.

The highest variability in milk productivity is observed in first heifers in the herds of "Savai" and "Gallyaaral" breeding farms, which amounted to 16.0 and 16.7%, respectively. In cows of the third and subsequent lactations it was in the range of 13.4-16%. The lowest coefficient of variability of milk productivity was observed in the group of bull-producing cows - 6.9-10.1%. It was within 8.3-13.1% in cows of the "Savai" breeding farm and 10.9-3.9% in the "Gallyaaral" breeding farm, which indicates a real possibility to increase milk yield in the herd as a whole.

The variability of milk yield level decreased with age, which, according to , is probably related to the selection process. Higher milk productivity of first-calf cows in the "Savai" breeding farm is associated with better preparation for calving and subsequent milking.

In "Savai" breeding farm preparation of heifers for calving starts from 6-months pregnancy. This gives the farm specialists an opportunity to reveal potential productivity of first-calf cows earlier and to conduct breeding work at a higher level, which leads to an increase in milk yields in the herd as a whole, creates conditions for a more reliable evaluation of bulls on the quality of offspring, reducing culling and increasing the period of use of cattle on farms.

In the herd of the breeding farm there are 25 first-calf cows with milk yield of 3200-4000 kg per lactation, as well as 17 cows with milk yield from 3900 to 5000 kg. This is the result of selection of first heifers, which were prepared for milking at the control yard. On the other hand, this is an additional evidence of high potential productivity of shvitzized cattle in hot climate conditions.

Fat-milk yield of cows. Average milk fat content indicators are at the same level in all farms of the region. They are lower than the standard requirements (3.9%) for Shvitz herd of Uzbekistan by 0.12-0.17%. Therefore, fat-milk yield remains the main breeding trait.

Live weight of animals. Live weight is one of the main traits in cattle breeding, positively correlates with milk yield. In all studied farms there is an increase in live weight of cows due to age and selection on this trait. The highest live weight of 474-494 kg had cows of 3rd calving and older, animals of breeding and bull-producing group.

In order to increase milk productivity and fat-milk yield, improve udder shape and suitability of animals for machine milking, the seed of bulls-producers was brought to farms from CSIO.Productivity of mothers of bulls-producers, the offspring of which served as material for the study, according to the data of breeding certificates ranges from 3660 to 7293 kg and mothers of fathers - from 4475 to 8746 kg of milk, the highest milk yield had mothers of bulls Apelsin 79 and Azot 2306, and the lowest - of bulls Vdovets 3517 and Otzav 6038. There are also significant differences in milk fat content. For the study we selected bulls-producers with high genetic potential of productivity, which can have a positive effect on the offspring

2.2 Methodology of immunogenetic studies.

As already mentioned, blood groups were determined in 2115

animals, including 791 daughters and 624 mothers. Some mothers, due to their retirement from the herd by the time the work started, were not studied. Another contingent of adult animals was studied, but by that time their offspring had either been sold or dropped out of the herd, which also prevented their blood types from being determined.

For determination of blood groups, 8-10 ml of blood was taken from each animal from the jugular vein. A solution of the following composition was used as a preservative: sodium citric acid (tri-substituted) - 32 g; glucose - 10 g; streptomycin - 500,000 units; discylated water I liter. Later, spermoyan (prescription) was used as antibiotics, which led to improvement of blood quality, lengthening of its storage time during transportation in conditions of elevated temperature.

Erythrocyte antigens of blood groups were determined in the laboratory of immunogenetics according to the generally accepted method of P.F.Sorokov, using a set of monospecific sera (reagents) of 60 names, produced in the laboratory of immunogenetics of the production and scientific association "Kuibyshevskoe" on selection in animal husbandry (Kuibyshev), which have passed a series of All-Union and International tests and meet the standards.

The following reagents were used in serologic reactions to detect antigenic composition for 10 blood group systems: A_1,$A_{2,ZB}{}^{(}{}_{1)(B)2}{}^{G}2^{G}3^{K}I$ IO1O2O3Ox P2Q,$T_{(1)}T_{(2)}$ $_{Y(1)}$ $Y_{(2)}A_{(1)}A_{(2)}$ $BDFF_3$ G I I $I_{(2)}$ O P(2) Q $P_{(2)}$ Q Y B G $C_{(1)}C_{(2)}ER_{(2)}R$ N $X_{(1)}G$ L F V $_{(1)}J_{(2)}$ L M $S_1S_{(2)}H$ $U_{(1)}$ U H Z

The reaction was performed in special plates with wells 8 mm deep and 5 mm in diameter with a flat bottom. The previously used blocks with hemispherical bottoms had the disadvantage that the reduction of the well volume led to a reduction in the area of the reacting media, which reduced their quality and prolonged the reaction time. In accordance with the generally accepted methodology, up to two drops of reagents were added to the horizontal rows of plates (blocks) with a special pipette. Up to I drop of 2.5% suspension of erythrocytes of the studied animals washed three times with physiological solution was placed in the vertical row. Then up to one drop of complement (rabbit blood serum) was dropped into each well, after shaking of the reacting components and two-hour incubation in the thermostat at +36-30o C the first "reading" of the reaction was carried out, and after repeated shaking and incubation the second "reading" followed. The results of the reaction were evaluated on a five-point scale: 0 - absence of hemolysis, 1-2-3 intermediate (partial) hemolysis, 4 - complete hemolysis. The study of blood groups of cattle was carried out according to the scheme below.

The study was carried out on the following groups of sexually mature animals: bulls - adult cows - their progeny. After determining the antigenic composition of erythrocytes in the animals under study, the examination of origin was performed. Then, blood group alleles of daughters, mothers and bulls of producers were determined on the basis of reliable progeny, after which the genotype of blood groups for all sex and age groups of animals was determined from the complex of identified alleles. In some imported bulls-producers the antigenic composition of blood groups was determined in VIZh (P.F.Sorokovoy) soon after their importation to the CSIO. The breeder bull Gnome 343 was retired from the herd before the beginning of the study, so his genotype was reconstructed by blood groups of his progeny and their reliable mothers: Gnome 343 A/-$G_3 0_t T_1$/-C_1W/WC'y/VL/-S_1H'0"/H'Z/-. Evaluation of bulls-producers on the quality of progeny was carried out after checking their daughters for the reliability of origin, which made the results of the study more accurate and objective.

Blood group research scheme

Plemkhoz Plemkhoz Plemkhoz

"Savay" "Izbaskan" "Galliaaral"

taking blood samples from animals, determination of antigenic

Use of blood groups in validation
origins

composition of animal erythrocytes, derivation of animal blood group alleles

study of the relationship between
blood groups and productivity

linear affiliation animals	reproducibly wild signs cows	level dairy products effectiveness	content fat in the mo locke

After establishing the alleles of blood groups of bulls-producers on the quality of offspring, their influence on the gene pool of herds of animals cultivated in the hot climate of Uzbekistan was determined.

The frequency of occurrence of antigenic factors was determined from the results of animal grading.

Methodology of statistical processing of the obtained results.

The frequency of antigenic factors was determined using the formula:

- $^{(R)}i = P/M$

-where ^{P}i is the frequency of antigenic factors,

-i is the number of animals with this antigen,

- m - total number of animals studied.

The frequency of occurrence of B and C alleles was determined by the formula;

$^{-P}i = 1i/2n$

- where P_i is the frequency of the allele,

- 1i is the number of animals having this allele,

- 2n the total number of animals studied.

The index of genetic similarity between the compared groups of animals was calculated according to the formula of L.A. Khivotovsky et al. /59/

- where g is the genetic similarity index,

- Pi, $_{Qi}$ - frequency of alleles.

To assess the productive qualities of carriers of certain allele genes, the average productivity of animals of these groups was compared with the herd average and with each other.

The relationship between markers - blood group genes and genes that determine economically useful traits of animals was determined by comparing the productivity of paternal half-siblings who inherited alternative paternal B-alleles.

Reliability of the difference between the mean values of productivity in the compared groups was determined by Student's t-criterion calculated by the formula:

$$t = \frac{M_1 - M_2}{\sqrt{m_1^2 + m_2^2}},$$

where m_1 , m_2 are the mean values of the trait,

Schr m_2 - statistical errors of mean values.

The following boning data were used:

- pedigree information,

- information on milk yield and fat yield,

- data on live weight at birth and during the subsequent period of post-embryonic development, which were taken from breeding cards of the form 2-MOL. Biometric data processing was carried out using the methods of population analysis described in the manuals of N.A. Plokhinsky and E.K. Merkuriea /123/.

III. RESEARCH RESULTS

3.1 Immunogenetic peculiarities of cattle bred in the studied breeding farms

Antigenic factors were determined using 60 reagents belonging to 10 immunogenetic systems: (A,B,C,F-V,J,L,M,S,Z,R'-S1).

After determination of antigenic composition and identification of alleles, the frequency of occurrence of antigenic factors in all studied animals was calculated. Some general regularities can be traced: antigenic factor F was the most frequent in all farms (breeding farm "Savai" - 0.9500, breeding farm "Izbaskan" - 0.9920, breeding farm "Galliaaral" - 0.9530). Further the sequence of farms in comparison remained the same: H' (0.8340; 0.9000; 0.9220), 0_x (0.9970; 0.9920:, 0.9910) and z (0.8040; 0.7470; 0.7500). Antigenic factors z'(0.0330; 0.0230; 0.0310),P_2(0.0830; 0.2030; 0.1090),$J_{(2)(}$0.0400;0.0840;0.0470) and X_1(0.02070; 0.0110; 0.0160)had low frequency of occurrence.

At the same time, there are certain differences on the analyzed issue. Thus, for example, in the breeding farm "Savay" antigens P_2, Y_1, B_1, B_2 E_3, B, and in the breeding farm "Izbaskan" - antigens 0_3, Y1, Y_2 in the herd of the breeding farm "Gallyaaral" were significantly more frequent O_1, O_2 Oz,$A_{(2)}Y_1$ Y_2 All these data indicate high genetic consolidation of animals of the same breed in herds of different regions of Uzbekistan.

3.2.Immunogenetic characterization of Schwyz breed bulls used for reproduction in the studied farms

For herd reproduction in the "Savai" breeding farm 12 bulls were used during one and a half years, in the "Izbaskan" breeding farm - 13. At the same time such bulls as Azot 2306, Antey 7438, Herb 9457, Patrick 2I37I5, Weet-Dawn 162483 and Shatun 3266 worked in dairy herds of both farms.

The bulls Shatun 3266 */132/* and Azot 2306 /86/ had the largest number of progeny in the breeding farm "Savai". The worst indicators are bull Otzav 6048 - only 2 heads and bull Elevat 170836 - 3 offspring. In the breeding farm "Izbaskan" most of all daughters of bulls Don 2575 and Antey 7438 - 80 and 70, respectively. Here the bulls Hazar 7757 and Vdovets 3517 had the fewest offspring.

In Schwyz breeding bulls used for herd reproduction, 16 alleles of the B-system were detected, in the "Izbaskan" breeding farm 18 different alleles of the B locus were detected. In both compared populations of bulls 14 alleles of 0-system locus were detected. Quite a high level of hereditary polymophism of blood group alleles revealed in the studied farms opens wide opportunities.

3 bulls with BG3O1T1 and BB allele were found in the Savai breeding farm. Other alleles of B-system are found in no more than 1-2 bulls.In C-system, allele $c^{R}2$ was found in 3 bulls, allele $^{C(C)}1^{(EW)}$in 3 bulls and allele $c^{"c"}$- in five bulls.

The most frequently occurring alleles B BG3O1T1 and $c^{i"i}$ were found in only three individuals in the herd of bulls-producers of the "Izbaskan" breeding farm, and the other identified alleles of the system were found in one or two animals. The allele c^{CEW} was found in 4 bulls, alleles $c^{c}2$, c^{w} and $c^{(mC}$)" - in three.

Having established immunogenetic peculiarities of the maternal herd in the breeding farms "Savai" and "Izbaskan", the antigenic composition of the herd of their daughters was investigated. Using father-mother-offspring triads, conducted an examination of the reliability of the origin of young stock. Determination of genetic features of the herd of offspring and their mothers made it possible to establish the change in the gene pool of the cow population as a result of crossing them with bulls-producers of the Schwyz breed.

A total of 40 blood group alleles have been identified in the B-system at the Savai breeding farm, which differ significantly in frequency of occurrence. Thus, allele BG3O1T1 has frequency of occurrence 0.1233, allele B^c - 0.1453, c°1° - 0.0659, $B^{G}J^{\circ\, I(T)}I^{Y}2$ - 0.0557, $B^{G''}$ - 0,0524. Among the rare B-system alleles the following can be noted: $B^{G}2^{P}2^{E}2^{G'}$, $B^{G}3^{\circ}1^{T}1^{A}1$Q'B I-G'I'G', $BI^{(1)}O^{1}Y^{2}$OY, BA2BQ. Their frequency of occurrence in the population was only 0.0017, and the alleles, $B^{G}2^{Y}2^{E}$Q-0.0034.

The gene pool of B-system alleles in the Izbaskan breeding farm was less diverse and represented by 32 alleles. Such alleles as c^{B}, $B^{G}3^{\circ}1^{T}1$, $c^{\circ}1^{o'}$ (0.1901; 0.0904; and 0.0536, respectively) are the most frequent. At the same time, the frequency of such alleles as in $B^{B}2^{(AO)}1^{P}2^{Q}$, $B^{Y}2^{E}2^{O}$ and $B^{Y)2(Y}$ $B^{(G)(')I(')(G)}$ was 0.0057; 0.0047; 0.0023; and 0.0023, respectively.

Differences in the structure of the gene pool of maternal herds in the two studied herds can be explained by the use of different breeding bulls in the reproduction of young stock, which left different B alleles to their offspring.

As a result, the gene pool of blood groups according to the B system became different. The frequency of animals with alleles that were present in the genotype of the used bulls-producers increased in the populations, as well as progeny with genotypes that were not found before appeared. Thus, animals with 19 different alleles of the B system were eliminated from the herd of the Savai breeding farm, but new alleles appeared in it . Among the latter are such alleles as $B^{B}2^{P}2$, B^{G} $^{30}1^{T}1^{Y}2^{B}2^{G}$, $B^{I}1^{O}2^{A}1^{(E)}$ 2J1 $B^{(O)}1^{T}1^{E}2^{(O)}$. They are inherited from breeding bulls Antey 7438, Herb 9457, Elevate 170836 and Jay I7I253.

After the use of the above mentioned bulls in the "Izbaskan" breeding farm, the gene pool of blood loci changed significantly: animals with alleles $B^{B}2^{G}2^{KO}X^{T}1^{X}1^{B}$, BMI^ , $B^{B}2^{P}2^{Y}2^{E}2^{G'Y'}$ were eliminated. , $B^{B}2^{\circ}1^{P}2^{A}1^{G'}$ and others. The frequency of alleles such as $B^{B}2^{0}1^{Y}2^{A}2^{(GP)}2Q^{Y}$, INB$1_t$ $B^{B}2^{P}2^{(Y)}2^{E}2^{GYI}$, B^{B}2O1P2A$1_{(}^{G)'}$ and others has increased. This is also true for the alleles $B^{B}2^{G}2^{KB}2^{G'}$, $1B^{B}2^{P}2$ $^{, B(B)}$, $B^{(G)}3°1^{T}1$, $B^{G}3°1^{T}1^{Y}2$ and some others. The frequency of their occurrence in the herd was 0.0601; 0.0374;0.02145; 0.0982; 0.0674, respectively.

32 different alleles of blood groups were detected in daughters of used bulls in the Savai breeding farm. Their frequency of occurrence is different and ranges from 0.0034 to 0.1368. It is especially high for alleles $c^{G}i^{(EW}$) (0,1368), $C^{"0"}$ (0,1367), $C^{X}2$ (0,1031$^{)}$, C^{W} 0,0642), A $_{and}$ $C^{G}1^{(W)}$(0,0541). Animals with alleles $C^{G}1^{R}2$, $C^{G}1^{X}2$, $A^{G'}$ (0.0034$^{)}$were rarely found in the herd.

The herd of the breeding farm "Izbaskan" differed slightly from the herd of the breeding farm "Savai" by system and included 41 alleles. Animals with alleles $C^{"C}$ (0.2160); $c^{c}1^{(w)}$ $^{(}$0.1620); A (0.0657$^{)}$; $C^{(C)}1^{(W)}$ $^{(}$0.0610); $^{C(C)}1^{W}$ 0.0587) and c^{W} 0.0540) were the most frequent. The cumulative frequency of occurrence was 0.6174. The alleles $C^{G}1^{R}2^{W}$ (0.0023); $^{C(C)}1^{ER}2^{W}$ (0.0026); $^{o(E)}$ $^{(}$0.0023$^{)}$; $c^{R}2^{G}$ 0.0070) and some others were rare.

The differences in C-system blood group alleles between maternal herds in the two compared farms are quite significant. Cows of the "Sawai" breeding farm lack alleles $^{c(C)}1WE$, $c^{(c)}1^{(EH)}2^{C}$, $^{c(c)}i^{R}2^{L'}$, which are found in animals of the "Izbaskan" breeding farm. which are found in the animals of Izbaskan breeding farm. At the same time, the alleles $c^{G}1^{WI1,}$ $^{C(C)}1^{ER}2^{X}2$, gEXgL and some other alleles were not found in the dairy herd of Izbaskan breeding farm.

After the participation of Schwyz breed bulls in the reproduction of the herd of the Savai breeding farm, the state of the gene pool did not change. Since they had no distinctive features from the structure of genetic maternal herd, no new alleles appeared in the population.

After the use of breeding bulls in the breeding herd of the Savai breeding farm, the gene pool of the C-system of blood groups was represented by 26 different alleles. Six of them from the herd eleminated $^{c}1^{P}2$: $C^{(C)}1^{WL}$:$C^{C}1^{ER}2$:$C^{(C)}1^{EL}$:$C^{(EWL)(:)}C^{R}2^{C}$ The frequency of occurrence of alleles $C°1^{(E)}$(0,2233), 0 "C" (0,1695), A (0,1100), A (0,0906) increased. This phenomenon is due to the use of sire bulls such as Antaeus 7438, Jay I7I253, Hall 172138, Testimonial 6038, West Lawn 162493 and Shatoon 3266, which left a significant number of their daughters in the herd. The frequency of the following alleles decreased significantly: $C^{e}1^{w}2$ (0.0338-0.0067); C(0)!^ (0.0405-0.0185); (0.0287-0.0019).

In turn, the gene pool of daughters of the Izbaskan breeding farm is represented by 38 different alleles. Six of them were eliminated from the herd: C°A' ' ; $_{C}$W ; $C°1^{Y}$-' . cA'A ; $A^{1'}$ and A . For the first time used breeding bulls, such as Herb 9452, Credit 7830, Khazar 7757, introduced into the gene pool of the daughter herd new, previously not found alleles of the C-system: $^{C(C)}1^{R}2^{WX'}$ (0.0068$^{)}$; $C^{G}2^{R}2^{C'}$ (0.0097); c^{G}1R1x2 (0.0010$^{)}$. The frequency of $c^{c}1^{2}$EW alleles increased (0.1620-0.2016); $c^{R}2$ (0.0017-0.0381); C^{W} (0.0540-0.1010$^{)}$. The number of animals with alleles $C^{C}1^{er}2$ (0.0188-0.0068); $C^{c}1^{r}2$ (0.0188- 0.0029); $Cc1^{C'}$ (0.0188-0.0048) and some others was decreased.

It should be noted that the gene pool of the daughter herd of "Izbaskan" breeding farm is much more diverse than that of "Sawai" breeding farm. In the herd of daughters of "Savai" breeding farm there are 5 alleles of B-system, which are not found in the daughter herd of "Izbaskan" breeding farm. In turn, the daughters of Shvitz bulls of "Izbaskan" breeding farm had 16 alleles of C-system in their gene pool, which are absent in the daughter herd of "Sawai" breeding farm. This can be explained only by the difference of the gene pool of the herd of bulls-producers and the mother herd. In case of wide use of bulls-producers in the conditions of industrial cattle breeding it is their genotype that has the most appreciable influence on the genetic structure of the offspring. When studying the herd of both breeding plants, practically no difference in paternal and maternal influence on the structure was revealed. At the same time, judging by the data obtained on the herds of breeding farms "Savay" and "Izbaskan", although the same bulls were used here (Azot 2306, West-Lawn 162493, Antey 7438, Patrick 2I37I9, Shatun 3266 and Gerb 9457), mothers had more influence on the structure of their gene pool here.

In addition to differences in blood groups between the populations of animals of the two farms under study, inter-lineage differences among the progeny of certain bulls-producers were also revealed. Thus, 43 different alleles of the B-system are present in the herd of the Savay breeding farm, and 39 alleles are present in the gene pool of the Izbaskan breeding farm herd. Significant diversity of the gene pool of breeding stock is indicated by the fact that in the herd of one breeding farm there are alleles of blood groups that are absent in the herd of another breeding farm, or vice versa.

Thus, the results of our research allowed us to identify genetic peculiarities of the use of Schwyz bulls-producers both on the mother herd and on the herd of their daughters. These features are manifested in the presence or absence of certain marker genes of B and C-system in animals. The revealed marker alleles of blood groups can be applied in creation of new intrinsic dairy types of cattle in conditions of Uzbekistan. Selection of animals by productive, as well as by marker alleles of blood groups typical for Shvitz cattle will allow to maintain a certain gene pool structure from year to year, which will be optimal for the given conditions of maintenance and breeding and guarantee maximum productivity of the herd.

3.3.Genetic characterization of Schwyz cattle lines

One of the interesting characteristics of cattle herds is their linear structure, as this is the main result of breeding work.

In this regard, genetic analysis of the progeny of animals belonging to different lines was carried out.

The data characterizing the features of genetic structure of B-alleles depending on the lineage of animals of the breeding farms "Savai" and "Izbaskan" belong to ten lines. In the herd of cows of the first farm separate genes of B-system are found only in one line, for example, $^{B(B)}2^{G}2^{KO'}xT1^{A}1^{B'}$- only in animals of Meridian 90827 line; $^{B(B)}2^{G}2^{K0}1^{O'}$- in Master 106902 and Enkel,-in the Master 106902 line; $BG_3O_{(1)}T_{(1)}Y_2E2G$" OF THE Meridian 90827 line; $c^{i}1^{0}1^{Y}2O'Y$ in the American Group line; BO $^{G'I(')(G)'}$-in the Master 106902 line. Other allelic genes are found in animals of many lines, such as $^{c(c)}$; $B^{G}3^{o}$ 1T1 : $B^{G}3^{O}Y^{1}$; $B^{I}1^{Y}2^{G}$: $B^{Y}2^{D'E(')}2^{O}$ others. As for C-system, anologous results were obtained here as well. The allele $^{c(c)}1^{E}$ is found only in animals of the Oregon line 0866, $c^{(}_{c)}1^{k}2^{ta}2$ ". in the EMO line YASH = 260, $_{c}ER_2WX2$ in the Butler line 107506, A - in the Master line 106902. In turn, such alleles as W , $C^{G}1^{R}1$, $C^{G}1^{WL}$,$c^{(c)}1^{Ex}2$ $c^{c}1^{ER}2$ $c^{c}1^{EC}$ and a number of others were found in animals belonging to many lines. The frequency of occurrence of allele genes is also uneven. For allele $B^{G}3°$1T1 in the Savai breeding farm it is 0.2609, for allele c^{B} - 0.0097. The same applies to other alleles.

The identified patterns arose because many bulls were used in different ways, some of them have many progeny and others less. On the other hand, the reason is also that many bulls-producers are characterized by the same B and 0 alleles, which are constantly inherited by the herds where they work. This leads to the leveling of alleles between lines.

A similar result was obtained in the Izbaskan breeding farm. Here alleles $B^{B}2^{G}2^{(KO)}X^{T}1^{A}1^{B}$ are found (0,0153) only in the Meridian 90827 line; $B^{B}2^{\circ}1^{Y}2^{A}2^{G,P}2^{Y'}$ (0,0148) - in the Master 106902 line; $^{B(B)}2^{X}1^{P}2$ (o.0097) in the American Group animal line; $^{B(B)}2^{\circ}1^{(P)}2^{A}1^{G'}$

(0.0129) in the line of the Azure, etc.д. In turn, such alleles as ; $^{c(c)}$; $B^{(G)}3°1^{T}1$; c1; c2 and others are characteristic of animals of many lines. In the C system, alleles $c°1^{B}$ (0.0633) are found only in animals of the Oregon line 086356; $C^{(C)}1^{(WX)}2$ (0.0391) and $C^{C}A^{o'}$ (0.0010) are from the American Troup lines; $0^{G}1^{R}2^{lE£}$ 2 (0.0068) is from the JAMO line JB=260. The alleles $C^{c}1^{(fi)}2^{w}$; $C^{G}1^{WL''}$; $C°1^{G}2$; $C^{(C)}1^{ER}2$; $C^{C}1^{EU}$ and others were found in animals of almost all lines.

This is also the result of using bulls-producers of different lines, but having the same alleles. And in the first case they worked in a herd of cows of the same line, which increased the frequency of occurrence of these alleles.

The following results were obtained during comparative characterization of animal lines bred in farms. For some alleles the difference in frequency of occurrence is practically not observed. Thus, the frequency of occurrence of $B^{B}2^{G}2^{KO}X^{T}1^{A}I^{B}$ in the "Savai" and "Izbaskan" breeding farms, respectively, is 0.0188 and 0.0153 (in the Meridian 90827 line); $B^{G}2^{Y}2^{E}2^{Q}$ in the same line) - 0.0129 and 0.0187. The difference was found for such alleles as $B^{c}2°1$(0.1128 and 0.0321) in the Master 106902 line, $B^{B}2(^{P)}2$(0.0100 and 0.0374) in the Azutchik line.For the C-system, the results are similar. In the Lazutchik line, the frequency of occurrence of the allele with $c^{G}1^{W}$ was 0.0605 and 0.0653, $0^{(CWL}$)' in the EMO line YASH-260 was 0.0432 and 0.0399, respectively. Some difference was observed for other alleles.

At the same time, we pay attention to the fact that some alleles are found only in animals of this or that farm, giving the lines certain characteristic features. This is a consequence of the use of different bulls in these farms. However, such bulls as Antey 7438 and Shatun 3266 and others worked in their herds, which gave the lines of both farms some homogeneity.In general, in both farms there is a significant difference between the lines on allele genes B and C-systems.

3.4.Immunogenetic similarities of Schwyz lines and flocks

Immunogenetic similarity of herds in experimental farms was estimated by the index of genetic similarity. The index of genetic similarity between the herds of "Izbaskan" and "Galliaaral" breeding farms had almost the same value, while in "Savai" breeding farm it differed significantly from the above-mentioned farms. When determining the genetic similarity between the lines of bulls-producers in separate farms, the lines with the most numerous offspring were taken as a basis. Thus, in the herd of the breeding farm "Savai" out of 12 analyzed lines of bulls only 6 lines were used for calculations. The least genetic similarity was noted between the progeny of the lines of Azot 2306 and Shatun 3266 bulls. It should be noted that in the herd of the breeding farm "Izbaskan", where the descendants of only 6 lines out of 13 were also left in the calculation, the difference between them was more pronounced than in the herd of the breeding farm "Savai". The interval of genetic index values ranged from 0.7760 to 0.9130.

Genetic equalization of animal lines in the "Savai" breeding farm may be the result of better selection and pedigree work, as in principle it is focused on a strictly defined economically useful trait - milk productivity.

For a clearer genetic separation of lines it is advisable to use blood groups as immunogenetic markers in this case directed crossbreeding in the selection of pairs will increase the level of heterozygosity in the offspring, which should positively affect their productivity.

In the herd of "Galliaaral" breeding farm only two out of 18 analyzed lines turned out to be of high value, so the indicators of genetic indices were not studied in this farm.

3.5.Verification of origin using blood groups

Immunogenetic control in the creation of new and improvement of existing breeds of cattle is necessary element in selection and breeding work.

According to experts, most often errors in filling out pedigrees occur for the following reasons:

-participation in insemination of two or more breeding bulls;

-replacement of calves at birth;

-unforeseen or intentional re-covering of uteruses bulls;

-lack of reliable animal numbering technology;

-errors in entering entries in the breeding book;

-Substitution or erroneous application of seed dosage

Taking into account the fundamental impossibility to carry out in-depth breeding without clearly established control over the pedigree of animals, one of the tasks of our study, as already mentioned above, was to check all young stock for the reliability of origin The reliability of origin was carried out according to the scheme: father-mother-descendant. At the same time, the analysis was based on the fact that the offspring should have hereditary factors that their parents had.

For this purpose, 383; 408 and 172 progeny of different bulls-producers were investigated in the breeding farms "Savay" and "Izbaskan" and the breeding farm "Galliaaral", respectively.In the herd of the breeding farm "Savay" the correct record in the pedigree was established only in 333 out of 383 progeny, i.e. the reliability of origin was 86.9%. In the breeding farm "Izbaskan" the correct records in the pedigree were established in 304 progeny, false in 104, or the reliability of origin here was lower than 74.5%.

In the "Gallyaaral" breeding farm, correct record was detected in 58 progeny, false - in 114, i.e. the reliability of origin was 33.7%. The data show that the level of zootechnical record keeping in the first two studied farms is at a rather high level.

According to the results of the examination of animals with reliable origin, the genotype of blood groups of all bulls-producers participating in herd reproduction in all three farms was established. Knowing it, as well as using the records of the young stock journal, we were able to identify true fathers in some offspring. Thus, in the "Savai" breeding farm it was done in 33 offspring, in the "Izbaskan" breeding farm - in 79 offspring, in the "Galliaaral" breeding farm - in 75 offspring.

As a result, the specific weight of animals with reliable origin increased: in the "Savai" breeding farm - up to 95.5%, in the "Izbaskan" breeding farm - up to 94.1% and in the "Gallyaaral" breeding farm - up to 77.3%

On the basis of our studies, appropriate changes were made in pedigree records, and all false descendants were excluded from further calculation and breeding schemes. As the studies showed, in each farm a sufficiently large number of bulls-producers were used, whose genotypes were characterized by a significant diversity, which played a positive role in establishing the true paternity.In general, for all three farms, the reliability of the origin of progeny was 72.1%.

3.6.Dairy productivity of animals due to genotype differences in blood groups

To reveal the influence of labeled hereditary factors of blood groups on milk productivity of offspring in the breeding farm "Savay" and "Izbaskan" the animals obtained from the following bulls-producers were studied: Azot 2306, Antey 7438, Gerb 9457, Shatun 3266.

Daughters of the bull Shatun 3266, who were carriers of the allele in $B^{G}3^{O}1^{T}1$, were of the greatest interest in this aspect. Their milk yield was 3520 ± 82 kg. The lowest result (3100 ± 89 kg) was in daughters of the bull Azot 2306, carriers of the allele $B^{(Y)}2^{(G)(')I'G}$. These data refer to the "Sawai" breeding farm. In the "Izbaskan" breeding farm, the highest and the lowest milk productivity (3418 ± 98 and 3117 ± 118 kg, respectively) had daughters from bulls with the same genotypes. The highest difference in milk yield among daughters of one bull carrying alternative alleles was found in the progeny of Herb 9457 (Savai breeding farm). Daughters-carriers of allele $B^{\circ}1^{T}1^{E}3^{\circ}$ gave milk 3105± 108 kg, and carriers of allele $^{c(c)}$ - 3317 ± *120, the* difference was 212 kg. In the "Izbaskan" breeding farm, a special contrast in milk yield was noted in carriers of the $B^{O}1^{T}1^{T}1^{E}3O$ allele B b . The difference was 382 kg with an advantage in favor of the Bb allele.

In the "Sawai" breeding farm the marked tendency on maximum and minimum milk yield retained its significance: both daughters and mothers with genotype $B^{G}3^{O}1^{T}1$ had the best milk productivity in the line of bull Shatun 3266, and carriers of allele in $B^{V}2^{G'I(')(G)}$had the worst index. It should also be noted that the level of milk productivity in mothers was lower compared to the level of milk yield of daughters of the same genotypes.

In the "Izbaskan" breeding farm, the descendants of the bull Shatun 3266 had the highest productivity - 3418+98 kg, and the lowest milk yield was observed in the descendants of Herb 9457 - 3005+95 kg. In contrast to animals of the herd of the breeding farm "Savai", milk productivity of daughters carrying allele $B^{Y}2^{G'I(')(')(G)}$in the line of bull Azot 2306 and cows with allele $B^{I}2^{(G)(')(')(G)}$ in the line of bull Antey 7438 was lower than productivity of mothers with the same genotype by blood groups.

The distribution of alleles of blood groups due to the difference in level of milk productivity in the herd of the breeding farm "Savai" from it, with the average value of milk productivity of 3273 kg, the spread of variability was 600 kg. It should be noted that for the herd of cows of both studied farms the set of "+" and "-" alleles in general coincided with the similar distribution of alleles for bulls, i.e. the highest milk productivity had animals with alleles $B^{G}3^{\circ}1^{T}$, $c^{c}2^{P}2$ and the lowest - alleles $b^{Y}2^{G\,1\,G}$and $B^{\circ\,1\,(T)}1^{E}\,3O'$.

Both in the herd of the breeding farm "Savai" and in the herd of the breeding farm "Izbaskan", the variability of the level of milk productivity in the C-system was much less than in the distribution. Animals with alleles $G^{G}1^{WL}$ had the highest milk productivity in C-system and the lowest - with alleles $c^{x}2$; c^{WL} . In whole animals with the most common alleles for B and C-systems had productivity close to the average value for the herd.Due to insignificant variability of fat level in milk in cows with different genotypes to blood groups, their distribution into "+" and "-" alleles was not carried out.

3.7.Reproductive traits of animals due to the difference of their genotypes by blood group loci

To investigate this issue, animals with reliable origin on both farms were divided into classes depending on the number of heterozygous loci in the genotype. It should be noted that there were no animals with 9 heterozygous loci in "Izbaskan" breeding farm. Cows with 8 and 9 heterozygous loci in the genotype had a high percentage of fertilization in "Sawai" breeding farm. 60% of cows with 8 and 9 heterozygous loci had a high percentage of fertilization. Low fertilization rate was observed in cows with 1 heterozygous locus - 31.2%. The animals with 5 heterozygous loci in their genotype were the most in the herd, their fertilization rate was 47,7%. Animals with 1 heterozygous locus 38,6% were not fertilized and came to repeated hunting, and the lowest indicator was observed in cows with 8 heterozygous loci in their genotype 16%.

It should be noted that the proportion of animals that came to re-hunt was practically among all cows having from 6 to 9 loci in the genotype.

In the "Izbaskan" breeding farm, the fertilization rate of cows was lower and differences in this indicator due to the state of genotype by blood groups were less pronounced. Thus, among the animals with 1-5 heterozygous loci in the genotype, the fertilization rate varied from 18.2 to 25.5%. Only starting from the group of cows with 6 heterozygous loci in their genotype, this indicator began to increase to a maximum value of 55.5%. The difference in the number of animals that came to repeated hunting was also less significant. But in the fertility of cows depending on the presence in their genotype of heterozygous loci for 10 blood group systems in general, the revealed trend remained: it was more characteristic for cows with 1 and 2 heterozygous loci than for animals with a larger number of the same loci

3.8.Economic efficiency of immunogenetic research implementation in the practice of breeding livestock breeding

Economic efficiency of livestock breeding in the studied farms is achieved by a complex of measures, namely further optimization of feeding, housing and the level of breeding work.

A great reserve is the use of animals with high genetic potential on the background of full-fledged balanced rations. The genetic potential of animals, in turn, is achieved by a high level of breeding work based on the use of achievements of many sciences: biology, genetics and immunogenetics.The economic efficiency of immunogenetics use consists of two factors: determining the reliability of the origin of animals and selecting them for desirable alleles.

The VNGO for breeding in livestock production approved a formula for calculating efficiency based on the results of testing young stock for provenance :

E = (K-N) (Ts-p-Cm - (C N),

where: E - economic effect,

K is the number of animals examined to control the validity of the origin,

H is the number of animals with false origin,

Price - cost of animals when sold for breeding,

Cm value of animals when sold for meat,

C is the cost of research per head,

N - number of all animals (offspring and their mothers) studied.

We have determined the economic efficiency of selection of animals by immunogenetic markers. Animals with these alleles can be used for inbreeding to create a type of cattle with high productivity potential. To calculate the economic efficiency of this measure we used the formula:

E = TS S P L - K,

Where Ts - purchase price of a unit of production, rubles,

C - average animal productivity, kg,

P - average gain of main production, expressed in % per head of animals of new type in comparison with initial ones,

L is a constant reduction factor for the results associated with the additional costs of surplus production,

K - number of animals of a new type or improved breeding achievement.

Cows with alleles desirable for these farms additionally gave respectively 235 and 169 kg of milk per head.

IV. DISCUSSION OF THE RESULTS OBTAINED

The study of erythrocyte antigen polymorphism in farm animals has expanded the possibilities of zootechnical science. For the first time it became possible to objectively evaluate genetic processes in animal populations during the formation of new breeds and breed groups. At present, it is possible to quantitatively characterize such phenomena as different types of crossbreeding leading to the phenomenon of heterosis using blood groups.

Recently, in order to increase milk and meat productivity in our country, large-scale blood transfusion of imported animals to local, well acclimatized breeds of cattle has been applied.

Intensification of livestock breeding has led to a radical decrease in animal diversity and the reduction of existing cattle frontiers. Striving for standardization of housing and feeding conditions reduce diversity within breeds and lines.

This process can be observed not only at phenotypic but also at genotypic level. Thus, in our studies, the use of imported Schwyz bulls resulted in a decrease in the number of B-locus alleles from 40 to 25 in the Savai breeding farm and from 33 *to* 30 alleles in the Izbaskan breeding farm. It is quite possible that these indicators depend on the intensity of use of bulls-producers, as well as their individual genetic features. Besides, as we have already mentioned, several of the same bulls were used in both farms. The process of changing the gene pool as a result of using Schwyz sires consists not only in the decrease in the number of blood group alleles, but also in the disappearance or appearance of a number of them.

Thus, for example, in the "Sawai" breeding farm such B-system alleles as $B^{B}2^{G}2^{KO}1^{(TA)(B)}$; $B^{B}2^{G}3^{(KO)}1^{O'}2^{(KO)B(B)}2^{I}1^{P}2$ $^{B}(I')1^{(O)}2^{A}1^{E}2$ $^{J}2$ $B(^{T)}2^{B}Q$ At the same time, others appeared: $^{B(B)}2^{(P)}2$, $B^{I}1^{(O)}2^{(A)}1^{(E)}2$ $^{J}2$ The same B system alleles were eliminated in the herd of Izbaskan breeding farm. But the set of appeared alleles turned out to be somewhat different;. $^{C(C)}2^{X}1^{(P)}2$; $B^{(B)}2^{P}2^{Y}2^{E}2^{G'Y'}$ $C^{B}2^{O}1^{P}2^{A}2^{G}$. $_{B}G_{3}T_{2}I_{2}Ap'G'G''$ $_{B}T_{(2)}B$ "Q

In wild populations of freely crossbreeding animals, such changes are probably related to differences in the survival rate of carriers of certain alleles. In populations of farm animals, these fluctuations can be explained by a possible connection of certain blood groups with some productive traits, since selection of livestock with a certain genotype has never been carried out in the studied farms.

Analyzing the process of decreasing the set of alleles of B and C-systems of blood groups as a result of selection, it can be assumed that such a phenomenon will occur in the future. This, according to some authors, may lead to a decrease in the genetic potential of animals, to an increase in homozygosity, and then to a decrease in a number of productive traits.

Antigenic factors of cattle blood can be used to determine genetic distances between different breeds and lines of animals. Since six common bulls were actively used in the studied breeding farms, no reliable differences in the frequency of blood group alleles between the groups of animals of both farms were found. The difference of some bulls-producers by blood groups allows to use these or those alleles as markers in selection on lines and families.

The question of the relationship between blood groups and certain productive traits was solved by studying the differences in milk productivity in animals with different genotypes for erythrocyte antigens. Thus, we found a positive correlation with milk yield in cows carrying the $B^{G}3^{\circ}1^{T}1$ allele, and a negative correlation with milk yield in cows carrying the $B^{Y}2^{G''I''(G)}$allele. The difference in milk yield between groups of animals with these genotypes was more than 400 kg. Contradictory results were obtained for other blood group alleles.

As already mentioned, fat-milk yield is a trait with less significant variability. The highest value of fatness was observed in cows carrying the allele c°1°-3.99% (Izbaskan breeding farm). In the "Savai" breeding farm, the differences in fat yield between groups of animals with different genotypes were not more than 0.1%.The difference of genotypes by blood groups also affected reproductive functions. To study this trait, animals were divided into groups according to the number of heterozygous loci. The most pronounced relationship between the number of the latter and fertility was expressed in the Savai breeding farm: the more heterozygous loci were in the genotype, the higher was this index and the lower was the number of animals that came to repeated hunting.

Such an approach to the analysis of quantitative traits in connection with the peculiarities of the genotype structure, in the opinion of many researchers, is more effective than comparing animals by one allele, as it reflects polygonal inheritance.

The advantage of heterozygotes for blood groups with respect to milk productivity and fat content has been pointed out by many authors. P.F.Sorokovoy /134/, A.M.Mashurov /92/ and some others emphasize the importance of uniform or close to such distribution of homo- or heterozygous loci, which will represent the optimal genotypic structure and favorably influence milk yield and fat content.

Regarding the reproductive abilities of cattle in connection with their differences in blood group loci, the following can be said. Many authors /60, 108, 132/ and before pointed out the increase of reproductive qualities at increase of dissimilarity of animals on loci of blood groups. As it is supposed, this difference has a positive effect on the embryonic development of individuals.

The discussion of the importance of blood groups in controlling the reliability of origin deserves special attention. Currently, this is the only objective and most accurate method that helps not only to find errors in breeding records, but also to reliably confirm the origin of a particular animal: B.A.Bagrii et al. /7/, Z.N.Vagonis et al. /21/, A.S.Vsevolkikh /38/, V.G.Gorin /51/, VLopushkova /84/.

As a result of our research, the relevance of the use of immunogenetic methods for cattle breeding was confirmed. In addition, significant discrepancies in the records of pedigrees of animals, which ranged from 13.1 to 66.3%, were revealed, and the genotypes of all imported bulls-producers, including those that have already been retired from the herd, were established;

V.Conclusions

Our findings allow us to draw the following conclusions:

1. The gene pool and polymorphism of herds by blood groups were revealed in the leading breeding farms of Uzbekistan, where shvizized cattle are selected. In total, 60 erythrocyte antigens were detected in the studied animals, the frequency of which ranged from 0.033 - z to 0.950 - f . Forty-nine alleles in the B-system and 50 alleles in the C-system were identified with frequencies of occurrence of 0.0017-0.2045 and 0.0023-0.2283, respectively.

2.The gene pool of the herd when using the same sire bulls changed not unambiguously, so as a result of using bulls Azot 2306, Antey 7438, Gerb 9457 and Shatun 3266 the offspring not only had changed allele frequencies, but also eliminated some of them ($B^{B}2^{I}1^{P}2$; $B^{T}2^{B\,Q}$; $B^{T}2^{B}$ ^{Q} ; $^{B(B)}2^{G}2^{K0}1^{0\prime}$) and others appeared ($B^{B}2^{P}2$; $B^{I}1^{O}2^{A}1^{E}2$ $^{Y}2$ $B^{I}1^{T}1^{E}2$ $^{O)}$.

3.In some bulls-producers, such as Azot 2306, Antey 7438, Gerb 9457 and Shatun 3266, only characteristic alleles of the B-locus have been identified, they can be used as genetic markers of lines ($I_1Y_2G'G''$; C_2P2 ; $C_{(2)}G_3\,O_{(1)}T_{(1)}$).

4.High genotypic similarity on blood groups between separate herds and lines of Shvitz cattle of Uzbekistan was revealed. The herd of breeding farm "Savay" had an index of genetic similarity with the herd of breeding farm "Izbaskan" r = 0.8850, and with the herd of breeding farm "Gallyaaral" r = 0.8254. Between the herds of "Izbaskan" and "Gallyaaral" breeding farm the index of genetic similarity is x = 0.8156.The index of genetic similarity between the lines in the herd "Izbaskan" ranges from r = 0.7760 - 0.9130. 5.It was established that progeny - carriers of alleles $B^{G}3^{\circ}1^{T}1$; $^{in(c)}2^{P}2$, which inherit from used bulls of producers, showed positive and allele $B^{Y}2^{G_1I,(G)\prime}$ negative relationship with milk yield in both studied herds. Cows with these

alleles had milk yields of 3520, 3418 and 3100 kg, respectively.
Low milk productivity in both studied herds is characterized by cows carrying allele in $B^{O}1^{E}3O$ (3105 and 3005 kg). The difference of genotypes by blood groups in connection with the level of milk productivity in B-system was more significant than in C-system. 6.It was found that with increasing heterozygosity of the genotype of the animal there is an increase in their fertilization rate and a decrease in their coming to repeated hunting. Thus, animals heterozygous for 8-9 systems of blood groups were fertilized from the first insemination by 60% and 16-20% of heads came to hunt again, while in animals heterozygous for 1 system these indicators were equal to 31.2 and 38.6%, respectively (breeding farm "Savai"). Similar data confirm this in the "Izbaskan" breeding farm, they amounted to 55.5; 22.2; 19.2 and 42.3%.

7.When carrying out esperty of reliability of the origin of repair breeding animals in the herds of the studied farms, it was found that errors in breeding records ranged from 66.3% (Gallyaaral breeding farm) to 13.1% (Savai breeding farm), and for individual bulls-producers these indicators were in the range of 5.5-70.5%.
Using blood groups, it was possible to find true fathers in 65.7% of animals that had erroneous records in documents.

8.Selection of animals taking into account their genotype will allow, based on our data, to increase milk productivity in the studied herds by 169-235 kg

VI. Proposals to production

In order to increase the level of milk productivity of the herds of the studied farms, we propose to use immunogenetic tests along with the generally accepted zootechnical methods for selection of young animals. It is recommended to repair the herd with animals having alleles $^{G}3^{\circ\ 1(T)}1$ and $^{C}{}_{2}{}^{(C)(}{}_{2)}$in their genotype. Animals with alleles $Y_2G'i'G\ O_1T^{(}{}_{1)}E_3$ should be culled. It is advisable to make special selections of animals when fixing bulls in order to obtain as many progeny with the desired marker alleles as possible.

At selection to give preference to animals heterozygous for 7-9 systems of blood groups, which will increase the fertility of cows from the first insemination in two times and significantly reduce the percentage of mothers coming in repeated hunting.

LITERATURE

1. Aleksandrova G.M., Bakhmutova T.V. Immunogenetic evaluation of lines of Kostroma and Shvitzkaya breeds // Proceedings of the All-Union Agricultural Institute. - Vid. 2, 1977. - C.27
2. Anisimov N.F. The use of blood groups to establish the origin of animals and to clarify the results of bulls evaluation on offspring / Animal breeding. - № 9, 1972. - C.58-59.
3. Anisimov N.F. About prediction of parental pairs combinability by blood groups / Proceedings of Velikoluksky Agricultural Institute. - Issue 24, 1972. - C.15-17.
4. Antigenic polymorphism of erythrocytes in cattle and its application in breeding work Leshkauskas I., Tsalias A., Vagonis 3.// Proceedings of the Lithuanian Research Institute of Animal Husbandry. -T.P, 1972. - C.5-II.
5. Bagrii B.A., Meshcheryakov VL. Immunogenetic control in animal husbandry// Livestock breeding. -1 4, 1985. - C.36-37.
6. Baranov A.V. Immunogenetic markers of imported Shvitzky bulls and their use in the creation of intra-breed dairy type of cattle of the Kostroma breed // Dissertation of Candidate of Agricultural Sciences. - Kostroma, 1986. - 0.48.
7. Bakhmutova T.V. Evaluation of bulls genotype with the help of immunogenetic markers // Reports of VASKHNIL. - № 5, 1980. - C.43- 44.
8. Bakhmutova T.B. Blood groups of Shvitz cattle and their use in selection // Abstract of dissertation of candidate of agricultural sciences. - Dubrovtsy, 1981. - C .26.
9. Bernikova N.N. Characterization of the antigenic composition of erythrocytes of cattle breeding farm "Lesnoe" // Autorefer diss... Candidate of Biological Sciences. -Tartu, 1972. - C.20.
10. Bernikova N.N., Shadmanov S.I. Milk productivity in dependence

on the presence of alleles and zygosity of blood group systems in animals of the breeding farm "Lesnoe" / Theoretical bases of breeding, genetics and selection of agricultural animals.1979.

11. Proceedings of VNIIRZh. - L.: Pushkin - Vyp. 23, 1976. - C.27-30.
12. Borozdin E.K. Use of genetic markers in selection of farm animals // Theory and practice of breeding in animal breeding in Russia. - M., 1980. - P. II3-II9.
13. Borozdin E.K., Okhapkin S.K. Prizygotic selection in multiallelic loci in cattle // Genetics. - C.330-335
14. Borozdina A.P., Slepchenko A.R., Krivulko V.M. Some aspects of the use of immunogenetic indicators in the selection of Shvitz cattle // Agricultural Biology. - VOL. P. - NO. 2, 1976. - C.205-209.
15. Bukarov N.G. Biological markers in cattle breeding // Agriculture abroad. - №10, 1984. - C.43-47.
16. Bukarov NG, Sorokovoy H. Efficiency of genetic labeling of cattle by blood groups // Livestock breeding. - №7, 1987. - C.19-21.
17. Bykovchenko Yu.G. Study of genetic polymorphism of blood groups of Alatau breed of cattle and the possibility of its use in selection and selection / Abstract of dissertation.... Cand.biolog.nauk. - Dubrovitsy, 1969. - C.24.
18. Vagonis Z.I., Meskauskas C.P., Podenas B.E. Determination of paternity by blood groups // Dairy and beef cattle breeding. -№ II, 1966. - C.38-40.
19. Vagonis Z.I. Use of immunogenetic indicators in selection and breeding work with pedigree cattle //Proceedings of the Lithuanian Research Institute of Animal Husbandry. - T.14, 1976. - C.146-156
20. Vagonis Z.I., Meskauskas C.P., Podenas B.M. Determination of origin by blood groups // Dairy and beef cattle breeding. - № 6, 1966. - C.27-29.

21. Vagonis Z.I., Meshkauskas C.P./Podenas B.I. Use of blood group analysis method in Lithuanian breeding cattle breeding Livestock breeding. - NO. II, 1970. - C.57-59.

22. Vagonis Z.I., Vinikas A.A. Immunogenetics at the service of livestock // Bulletin of agricultural science. - NO. II, 1985. - C.103-110.

23. Vasilieva MA, Potalylova VG Influence of "blood infusion" of imported Schwyckian bulls to increase protein and fat content of milk of the Kostroma breed. Ways to improve breeding and productive qualities of cattle // Proceedings of VSKhIZO. M.,1983. C.30-38.

24. Vaher L. On the use of blood groups in the selection of pairs // Proceedings of Eeton.NII of animal breeding. - 1 31, 1973. - C.13-18.

25. Vaher L. 0 inheritability of blood group alleles of Estonian black-mottled bulls. Improving the efficiency of genetics and breeding methods in animal breeding // Abstracts of the scientific conference.

26. Conf.-Baisogola, 1978. - C.108-110.

27. Vinikas A.A. Use of blood groups to solve practical problems of breeding cattle breeding // Research works of the Lithuanian Research Institute of Animal Husbandry, completed in 1978. - Kaunas, 1979. -C.3-5.

28. Influence of bulls-producers on the formation of allelofond of lines Mashurov A.M., P.F.Sorokovoy, Dedov MD./.Increasing the efficiency of methods of genetics and breeding in animal husbandry // Theses of dokl.dokl.nauchnogo conf. - Baisogola, 1973. - C. 101-103.

29. The possibility of improving the technique of selection of bulls on the basis of immunogenetic studies /Bakovchenko Y.G., Yancheva R.G., Sorokovoy P.F., Nazarov U .S./. // Proceedings of the Kirg.

Research Institute of animal breeding and veterinary medicine. - Frunze" - Vyp. 26, 1976. - C.41-49.

30. Possibilities of using blood groups in selection of Sychiv Shvitz cattle. /Sorokovoy P.F., Chernushenko V.K., Turkovich K.A., Timoshenko V/. // Collection of scientific works of the Latvian Agricultural Institute. - Vyp. 168, 1979. - C.33-37.

31. Vorobyev E.G. Blood groups of Kholmogorsky and Shvitsky cattle and their use in breeding work // Author's thesis. Candidate of Biological Sciences. -Dubrovitsy, 1970. - C.22.

32. Vorobyev E.G., Sorokovoy P.F. Genetic structure of the herd // Livestock breeding. - № 3, 1981. - C.32-34.

33. Voronenko V.I. Allelofond of wood groups of red steppe and Angler cattle // Bulletin, scientific works of the All-Russian Research Institute of Animal Husbandry. - Vyp. 65, 1982. - C.58-60.

34. Voronenko V.I. Use of immunogenetic markers in crossbreeding Angler and red steppe cattle //Autorefer" diss.... Candidate of Agricultural Sciences. Dubroviny, 1983. C. 23. I

35. Vysyaskikh A.S. Shvitskaya breed and methods of its improvement. - Moscow: Rosselhozizizdat, 1970. - C 256.

36. Vysyaskikh A.S. Improvement of brown breeds in conditions of industrial technology of milk production // Methods of improvement of cattle of brown breeds. - Tula: Priokskoe book publishing house, 1976. - C.25-27.

37. Vsevolkikh A.S. Brown breeds of cattle. - M.: Kolos, 1981. - C. 271.

38. Vsevolkikh AS, Alexandrova GL, Bakhmutova T.V. Imunogenetic method of verifying the reliability of the origin of living animals and its role in breeding // Proceedings of the All-Union Institute of Plant Protection. - Vyp. 73. - C.7-12.

39. Vyaskikh A.S., Kuransky M.S. Imported cattle in the USSR. - M.:

Kolos, 1976. - C. 7-34.
40. Vyaskikh A.S., Aleksandrova GL, Bakhmutova T.V. The use of blood groups in the evaluation of bulls-breeders on the quality of offspring // Proceedings of the All-Union Agricultural Union. - Issue. 145, 1978. - C.17-23.
41. Vyaskikh A.S., Aleksandrova GL, Bakhmutova G.V. Use of immunogenetic methods in the selection of dairy cattle // Material of the XUT International Conference on blood groups and biochemical polymorphism of animals. - M. - T.SH, 1979. - C. 78-81.
42. Vyaskikh A.S., Sorokovoy P.F., Bakhmutova T.V. Immunogenetic control in breeding and selection of high-yielding cows // Breeding cows for dairy complexes. - I.; Kolos, 1981. - C.237
43. Breeding of new lines of Sychevsky and Shvitsky cattle with participation of genetic markers of B-system of blood groups (Mernushenko V.K., Kostin B.P., Turkovich K.A., etc.). Increasing the efficiency of genetics and selection methods in animal breeding // Theses of the report of the scientific conference -Baisogola, 1978.- C.98-100.
44. Genetic structure of cattle population by polymorphic systems in connection with methods of breeding and selection /Merkuryeva E.K.,Skripnichenko G.G.,Kryukova A.D.,Philipson V./. // Proceedings of the XY1 International Conference on blood groups and biochemical polymorphism of animals. - VOL. 6. - T., 1979. - C.I5I-I55.
45. Genetic polymorphism of blood groups in domestic cattle breeds /Sorokovoy P.F., Vorobyev E.G., Mailyan L.D., etc./ // Collection of scientific works of the All-Russian Research Institute of Animal Husbandry, - Issue 16, 1963, - P.8-15.
46. Genetic markers of breeder lines and their use in breeding / Vsyakikh A.S., Aleksandrova G.M., Sorokovoy P.F., Bakhmutova

T.V./ // Reports of VASHNIL.-MO, 1976. -C.28-32.
47. Genetic markers and milk productivity of Sachev cattle of new lines /Sorokovoy P.F.,Gontov M.E.,Chernushenko V.K.Grigoriev N.N./. // Livestock breeding .- No. 9, 1985, - P. 34-36.
48. Goncharenko G.M. Immunogenetic characterization of Altai red and Angler cattle for the purpose of its further improvement // Author's dissertation...kand.s.-kh.nauk.nauk.Novosibirsk, 1987. - C.18.
49. Gontov M.E. Milk productivity of Sychiv cattle in connection with the combination of marker B-alleles //Bull.nauchnykh nauchn.raboty VNII of animal breeding. -Вып.69,1983.-C.42-43.
50. Gorin V.G. Genetic control over the correctness of records on the origin of breeding animals // Breeding in animal breeding. -Minsk :Urozhay,1968.-S.69-73.
51. Grodula Y.D. Immunogenetic verification of the data of origin is an obligatory element of breeding work // Use of genetic methods in selection of agricultural animals.-Riga: Zvaigene,1973.- P.60-62.
52. Grinitsina G.A., Shadmanov S.I., Novikova D.N. Dairy productivity of cows at different combinations of alleles of blood groups // Scientific Proceedings of the Omsk Agricultural Institute named after S.M.Kirov.-T.160, 1977.-S.30.
53. Blood groups in Lithuanian breeds of cattle and prospects for their use in selection and breeding work (Vagonis Z.I., Meshkauskas C.P., Podenas B.I., Antanavachus D.I.) // Genetics and new methods of selection of dairy cattle breeds.-M. .'Kolos, 1970.-C. 424-431.
54. Blood groups as markers of genetic changes in cattle/Trelja J.,Trelja E.,Stavazh K.,Markovic L./.// Proceedings of the XY1 International Conference on Blood Groups and Biochemical Polymorphism of Animals.-T.2.-L.,1979.-0.139-144.
55. Deeva V.S. Characterization of Siberian black-breed cattle on

antigenic properties of blood and its use in breeding// Author's thesis...kand.s.-kh.nauk.Novosibirsk, 1986.-S.19.

56. Dexne V.Ya. Dairy performance of daughters of bulls-producers by inherited paternal alternative allele // Use of immunogenetic methods in pedigree animal breeding // Theses of the symposium. - Baisogola, 1976. C.14-16.

57. Zhebrovsky L.S., Mityutko V.E. Use of polymorphic protein systems in breeding. - L.:Kolos, 1979. - C.184.

58. Zhivotovsky L.A., Sorokovoy P.F., Maturov A.M. 0 calculation of genetic similarity indices between animal populations by frequencies of genes controlling polymorphic traits // Genetics. - T.9. - 4, 1973. - c 126-131.

59. Ivanov G.I. Reproductive ability of cows and blood groups // Livestock breeding. - > 2, 1985. - c.48-50.

60. Study of the effectiveness of the use of genetic markers tsri selection and selection in livestock / Klabukov P.G. Bezenko S.P., Sorokovoy P.F. and collection of scientific papers of the Research Institute of Animal Husbandry. - Vyp. 23, 1971. - c.103-106.

61. Study of genetic markers and the effectiveness of their use in animal breeding / Klabukov P.G., Sorokovoy P.F. Bezenko S.P. et al. 7. Breeding and feeding of farm animals // Proceedings of the All-Russian Research Institute of Animal Husbandry. - T.PX1U, 1972. - C.124-135

62. Study of the imunogenetic structure of an oto-selected bull population /Trelja E., Trela J., Richlis T., Kraszewska D./. // Proceedings of the XFI International Conference on Blood Groups and Biochemical Polymorphism of Animals. - T.P. - L., 1979. - P I45-150.

63. Ilyinsky AA, Parakhin II, Potopalova VG Experience in the use of Schwyz bulls in the herd of the breeding farm [y]Karavaevo [(i)] //

Proceedings of the All-Union Agricultural Institute. - Vyp. 130, 1977. - P.3-7.II7
64. Ilyi K.Ya. Studying the gene pool of cattle by blood groups and its use in breeding // Abstract of dissertation.... Candidate of Biological Sciences. -Tftu, 1971. - C.19.
65. Immunogenetic features of imported bulls of the Shvitz breed / Vsyakikh L.S., Sorokovoy P.F., Bakhmutova T.V., Semyonov, 1980. - C.20-21.
66. Johansen M., Rendel J., Graverg 0. Genetics and breeding of domestic animals. - M.: Kolos, 1970. - C.347.
67. Use of blood groups in breeding work with cattle /Vaher L., Ykva T., Madele M., Zijsbuk T./.
68. Use of immunogenetic methods in breeding livestock // Abstracts of the symposium. - Baisogola, 1976. - C.6-7.
69. Use of genetic markers in breeding of dairy cattle /Fotiadis F.V., Smirnov O.K., Sorokovoy P.F., et al. // Livestock breeding. - M 22, 1983. - C.37-38.
70. Istomin AA, Shadmanov S.I., Bernikova N.N. Dairy productivity of black and sperm cattle in connection with the evaluation of the genotype of animals by blood groups // Bulletin. VNIRGZH. All-Union. Research Institute of breeding and genetics of agricultural animals. - Issue. 24, 1977. - C.34-37.
71. Kasparov G.I. Genetic features of Caucasian brown cattle // Livestock breeding. - Л 8, 1983. - C.24-25.
72. Kaczmarek A., Dorynok 3. Influence of selection on the structure of blood groups of pedigree bulls of the black-breed in Poland // Proceedings of the XУ1 International Conference on blood groups and biochemical polymorphism of animals. - T.P.-L., 1979. - C.135-138.
73. Klabukov P.G., Sorokovoy P.F. Blood groups of dairy cattle and

prospects for their use in breeding // Genetics and new methods of selection of dairy cattle breeds. - M.:Kolos,1976. - C.412-419

74. Klabukov P.G. Bezenko S.P., Sorokovoy P.F. Application of immunogenetic methods in the evaluation of producers // Evaluation of producers on the quality of offspring. "M.: Kolos, 1972. - C.I9I-I93.

75. Kochetkova N.P. Genetic structure of blood groups in Shvitz cattle of Smolensk region // Bulletin of scientific works of the All-Russian Research Institute of Animal Husbandry. - Issue 45, 1975. - C.13-14.

76. Krasilnikova A.I. Dire breeds of cattle and their distribution in the countries of the world. - M.:Kolos, 1972. - C.55.

77. Krasota V.F., Lobanov V.T. Breeding of farm animals. - M.: Kolos, 1976. - C.4I6.

78. Krasota V.F., Sorokovoy P.F., Slepchenko A. Analysis of genetic structure of populations of red humpback cattle. // Vesnik s-kh nauki. № I, 1970. - C.66-70.

79. Kurinsky M.S. Imported Schwyck cattle and its use // Breeding selection and breeding on lines of brown cattle breeds. - M.: Rosselkhozizizdat, 1968. - C.224-235.

80. Kusmoddaev K.S., Issyk V.V.. Immunogenetic features of Alatau shvitzized cattle // Genetics and selection of agricultural animals, 1986. - C.156-160.

81. Kushner H.F. Prospects of genetics and animal breeding by 2000 // Animal breeding. № 3, 1971. - C.82-86.

82. Kushner H.F. Problems and tasks of animal immunogenetics. - M.: Nauka, 1972. - C.214-225.

83. Lakin G.F. Biometrics. Moscow: Higher School, 1980. - C.293.

84. Lapushkova V.V.. Evaluation of bulls-producers on the quality of offspring with the use of immunogenetic control. Candidate of

Agricultural Sciences. - M., 1971. - C.18.
85. Lebedev ML. Heterosis in dairy cattle breeding // Heterosis in cattle breeding. - L.: Kolos, 1968. - G.107-118.
86. Lebedev ML, Dmitriev NG, Prokhorenko PN Interbreeding in dairy cattle breeding. -L.: Kolos, 1976. - C .271.
87. Maylian L. Groups of Caucasian brown cattle and their application in breeding work // Avtorefdass...kand.biol.nauk.- Zfevan, 1971. C.22.
88. Maksimenko V.F. Change of gene structure of blood groups of Yaroslavl cattle in the process of selection // Bulletin of scientific works of the All-Russian Research Institute of Animal Husbandry. Issue 38, 1974. p. p. 102-103.
89. Malkovich V.M. Study of correlation of blood groups with animal productivity // Proceedings of Latvian Agricultural Institute. - Issue 74, 1975. p. p. 55-59.
90. Marin V.G., Yakovleva Yu.A. Genetic potential of Shvitz cattle // Livestock. № 10, 1987. - 0.26-27.
91. Matousek I. Blood groups of cattle. - Kiev: Urozhay, 1964. C.147.
92. Mashurov AL. Immunogenetic analysis of selection processes in populations of Simmental cattle // Authsref. diss.... Candidate of Biological Sciences. Dubrovitsy, 1972. - C.20.
93. Mashurov AL. Genetic markers in animal breeding. - M.: Nauka, 1980. - C.318.
94. Mashurov AL, Some regularities of internal organization of Simmental cattle herd of breeding farm "Elansky" by blood groups // Genetic bases of selection of cattle. - Kiev: Naukova Dumka, 1981. - C. 96-99.
95. Mashurov A.M. Cattle breeding using genetic markers // Livestock breeding. - й 4, 1984. - C.34-37.

96. Mashurov A.M." Dedov MD, Sorokovoy P.F. Some aspects of the use of blood groups in the analysis of selection processes in populations of Simmental cattle // Obs. of scientific works of the All-Russian Research Institute of Animal Husbandry. "Vop. 22, 1971. C. 12-17.
97. Mashurov A.M., Zhivotovskiy A.A. Structure of the herd of cattle by blood groups for selection // S-kh biology - T.12, 1977. - C .279-280.
98. Melkonyan L.M. Use of immunogenetic markers in the evaluation of bulls genotypes on the quality of offspring]/ Bill, scientific works of the All-Russian Research Institute of Animal Husbandry. - Issue. 69, 1983. - C.38- 41.
99. Merkuryeva E.K. Biometrics in breeding and genetics of agricultural animals. - M.: Kolos, 1970. - C.423.
100. Merkuryeva E.K. Genetic bases of selection in cattle breeding. - M.: Kolos, 1977. - C.240.
101. Methodical recommendations on the use of hereditary polymorphism in breeding work and selection-genetic research with cattle and pigs in Ukraine /Voron F.P., Marinchuk G.E., Meshcheryakov V.Y. and others/. - Kharkov, 1975. - C.87.
102. Methodical recommendations for research and use of immunogenetic and markers" for control of origin of cattle /Sorokovoy P.F./. - Dubrovitsy, 1981. - C.88.
103. Methodical instructions for studying the relationship of polymorphic blood systems with economically useful and other biochemical traits of cattle and their use in breeding / Zhadmanov S.I./. - Tashkent, 1981. - C.70.
104. Mescheryakov V.Y. Blood group and their use in breeding cattle // Heredity of economically useful traits in farm animals. - Kiev, 1963. - C.24-31.

105. Misostova N.V., Vininchuk D.T., Podoba B.E. Blood group system "B" alleles as markers in the study of the genotype of the bull sire // Agricultural Biology. - T.H , 1974. - C. 732-733.

106. Meat productivity of crossbred cattle under the influence of the combination of parental pairs on genetic and biochemical individuality /Smirnov O.K., Sorokovoy P.F., Bukarov N.G., et al. // S-kh biology. - T.XY1. - M 5" 1981. - C.661-664.

107. Nedava V.E., Podoba B.E., Tsiduiko G.A. Use of blood groups as markers for evaluation of genealogical selection of cattle populations // Genetic methods of evaluation of agricultural animals. - 1985. g C.II-I5.

108. On the use of blood groups and hereditary types of serum blood proteins in the selection of cattle / Vaher L., Ykva T.Y., Nadel I., et al. // Proceedings of the XU1 international conference on blood groups and biochemical polymorphism of animals. - T.P.L., 1979. - C.194-195.

109. Features of blood groups and polymorphic proteins of blood serum in Sychev cattle /Chernushenko V.K., Sorokovoy P.F., Budnikova A.V. and others/. //Use of immunogenetic methods in breeding livestock // Abstracts of symposium mind. - Baisogola, 1976. - C.24-25.

110. Ohapkin SK, Vorobyev EG, Khrunova AI Use of animal blood groups for heterogeneous selection of parental pairs and increase heterozygosity of offspring // Proceedings VNIIPLZh.-No. 2, 1981. - C.62-67.

111. Ohapkin S.K., Khrunova A.I. Some results of the study of the interaction of natural selection and selection on the model B locus of blood groups of cattle // S-Kh biology № 8, 1984, P. 3-8.

112. Ohapkin S.K., Khrunova A.I. On the possibility of using blood groups to predict milk productivity of cattle // S-kh biologiya. - № 6,

1985. - C. 46-49.

113. Ohapkin SK, Rozhkov YI, Khrunova AI Stabilizing selection and B-lotsus of blood groups of cattle. // S-kh biology. - M 5, 1987. - C.5.

114. Pavlichenko V.P. 1 blood groups and milk productivity *//S-kh* biology. -№ I, 1983. - C.II3-II7 /

115. Pavlichenko VP, Pepina GD, Shinkareva VP Gene pool of blood groups of Ayrshire cattle in some farms of the Black Earth // Improvement of methods of breeding work and technology of livestock production. " T. 395. - L.1 Pushkin, 1980. - C.62-68.

116. Pavlichenko V.P., Bernikova N.P. Prospects for the use of blood groups to increase the efficiency of selection work in dairy cattle breeding // Use of interior indicators in selection and breeding work. - Л., 1982. - C. 58-67.

117. Pavlov V. Prospect of development of brown Schwyck breed in Canada and USA // Dairy and beef cattle breeding. № 5, 1972. - C. 46-47.

118. Pakhtusova 3., Vorobyeva G., Koksharova A., Kondratieva T. Use of imported Shvitz bulls // Dairy and meat cattle breeding. - Ж 2, 1987. - C.30-31.

119. Pepina GD, Shadmanov S.I. Immunogenetic characterization of Ayrshire cattle lines and milk productivity of cows depending on inheritance of alternative paternal alleles of blood groups // Increasing the efficiency of genetics and selection methods in animal breeding Theses of Dok L. Conf. - Baisogola, 1978. - C.104-105.

120. Pepina GD, Pavlichenko VP The use of blood groups in improving selection and breeding work with Ayrshire cattle // Increasing the effectiveness of selection and breeding work in animal husbandry. - Ж . L., 1980, - p.113- 117.

121. Plan of selection and pedigree work with brown cattle breeds

in the USSR / edited by A.S. Vsevolkikh /. - M.: Kolos, 1972. - C.27I.

122. Breeding work with brown cattle /Solntsev K.M., Begunov A. Krasilnikova L.N., et al. // Livestock breeding. - Ж 8, 1978. - C.42-47.

123. Plokhinsky I.A. Manual on biometrics for zootechnicians. "M.: Kolos, 1969. - C.256.

124. Podoba V.E. Some aspects of the use of blood groups in the selection of cattle // Dokl. VASKHNIL. - Ж 7, 1976. - C 30-33.

125. Primakin I.N., Korshunova A.A. Methods of effective use of imported bulls of Schwyck breed in refreshing the blood of the Kostroma herd in the herd of plemsovkhoz "Kostromskaya" // Ways to improve breeding and productive qualities of cattle // Proceedings of VSKhIZO, 1983. - C. 52-56

126. Rendel J. Use of blood groups and types of proteins serum to identify similarities and differences between animal populations // S-abroad. - й 10, 1968. - C.20-24.

127. Ruzhevskiy A.B., Ruban Y.D., Berdnik P.P. Breeds of cattle. - M.: Kolos, 1980. - C.I52-I7I.

128. Samartsev M.I. Shvitzized cattle and its improvement // Livestock breeding. - №8, 1986. - C.38-40.

129. Semenov V.A., Sorokovoy P.F. Dependence of offspring productivity on inheritance of alternative alleles of bulls genotypes by blood groups // Genetic bases of selection of cattle. - Kiev: Paukova dumka, 1981. - C. 72-73.

130. Semenova E.I. Genetic-automatic processes in the systems of polymorphic proteins in cattle // Livestock. -Л 12, 1985. - C.22-25.

131. Improvement of methods of selection of cattle on the basis of genetic and other studies /Smirnov O.K., Klabukov P.G., Bezenko S.P. and others/. // Bulletin of scientific works of the All-Russian

Research Institute of Animal Husbandry. - Vyp.31, 1977. - C. 3-7.
132. Soldatov A.P., Kutuzov M.S., Ionova S.E. Reproductive abilities of Shvitz cattle // Livestock. - NO. II, 1986. - C.41-43.
133. Sorokovoy LF, Vorobyev EG, Maylyan AD, Anisimov NF, Zhkovchenko YG Genetic polymorphism of blood groups in domestic breeds of cattle // Collection of scientific works VIZh. - Issue. 16. -Dubrovitsy, 1969.
134. Sorokovoy P.F. Application of immunogenetic methods in line breeding // Improvement of brown breeds of calves. - Yerevan: Haystan, 1972. - C.137-143.
135. Sorokovoy *P.F.*, Slepchenko A.R. Control of origin by blood groups *JJ* Dairy and beef cattle breeding. -№ 12, 1969. -c.18-19.
136. Sorokovoy P.F., Kvitko A.S., Bykovchenko Y.G. Relation of dairy productivity with homo- and heterozygous loci of blood groups Agriculture of Kyrgyzstan. - .№ 7, 1970. - C.29-30.
137. Sorokovoy P.F., Mashurov AL. The possibility of establishing paternity in cattle by blood groups *JJ* Materials of the 2nd conference of young scientists on genetics and breeding of cattle. - Л., 1971. - C.5-6.
138. Sorokovoy P.F., Chernushenko V.K., Dudnikova A.V. Genetic structure of Sychevsk cattle // Livestock. -№6 1981. - C.49-50.
139. Sorokovoy P.F., Bukarov N.G. Evaluation of the combinability of hereditary factors by genetic markers // Animal breeding. -II, 1987. - C.22-23.
140. Stormont K. Blood groups and their importance in the improvement of dairy cattle // Agriculture abroad. - Я 6, 1968. - C.24-26.
141. Tikhonov V.N. Use of blood groups in animal breeding. - M.: Kolos, 1967. - C.390.
142. Tikhonov V.N. Problems of using immunogenetic indicators

and obtaining heterosis in animals Heterosis in animal breeding. L.: Kolos, 1968. - C. 22(8-247.
143. Htsring F., Messerschmidt X. Cattle breeds in the USA, Canada, South America // Guide to animal breeding. - M.: Kolos. - Vol.3, book 1, 1965. " C. 422- 426. '
144. Chernushenko V.K., Gontov M.E. Character of inheritance of genes - markers of blood groups by highly productive cows in the process of selection // Bill, scientific works of the All-Russian Research Institute of Animal Husbandry. - Vyp. 79, 1985. - C.58-60.
145. Shadmanov S.I. Immunogenetic control in breeding and pedigree work // Dairy and beef cattle breeding. - 1 9, 1974. - C.43-44. Shadmanov S.I. Use of blood groups in breeding and pedigree work with cattle Sb. of scientific works of VNIIRI. - VNIIRI. 25, 1977.
146. Shadmanov S.I., Koreshkov V.A. Some results of the study of blood groups of cattle *JJ* Collection of scientific works of the All-Russian Research Institute of Animal Husbandry. - Vop. 16, 1969. C.21.
147. Shadmanov S.I., Bernikova N.N. Blood groups of black-breed cattle and their importance in selection // Feeding and breeding of animals. - Odessa, 1972. - C.143-145.
148. Shadmanov S.I., Bernikova NN, Istomin AD. Study of the possibility of predicting milk productivity of cattle by BLOOD groups *//Republican* scientific meeting on genetics and breeding of agricultural animals // Abstracts - Alma-Ata, 1975. - C.44-45.
149. Shadmanov S.I., Pepina G., Kovalenko G. Prospects for the use of immunogenetic indicators for labeling lines and families of cattle. // Breeding and use of Ayrshire cattle in the RSSR. - Issue. 4. - Л., 1976. - C.107-110.
150. Shadmanov S.I., Pepina GD, Lozgacheva O.A. Study of the

possibility of using blood groups in the selection of cattle Ayrshirokoy breed // Use of immunoge- netic methods in breeding livestock // *Abstracts of the* symposium. - Baisogola, 1976. - c.8-9.

151. Stukowski I., Kovac D., Czontosz G. Effectiveness of different methods of origin control in offspring of cattle // Proceedings of the XУ1 International Conf. on blood groups and biochemical polymorphism of animals. - T.P. - L., 1979. - c. 37- 39.

152. Ykva Introduction of immunogenetic control of the reliability of the origin of animals // Livestock breeding. - II, 1984. - p.II-I2.

153. Ykva T., Nadel . On the control of data on the origin of cattle on the basis of genetic markers // Theoretical and practical questions of veterinary science. Tartu, 1978. - c.266-268.

154. Eydrigevich E.V., Lapkin M.M. Genetic examination of the origin of cattle and analysis of the causes of errors // Protein polymorphism and its use in breeding. - Odessa, 1974. - c.26-33.

155. Eydrigevich E.V., Lapkin M.M., Petrenko VD. Selection of pairs in cattle breeding with the use of immunogenetic tests // Ways of increasing productivity of cattle and birds. - Odessa 1978. - C.53-58.

156. Eydrigevich E.V., Raevskaya V.V. Exterior of farm animals. - M.: Kolos, 1978. - C.255.

157. Eydrigevich E.V., Petrenko V., Karnozhitsky V.. Polymorphic-& nd protein systems in animal breeding // Animal breeding. - Ж 7. - C. 28-29.

158. Eisner F.F. Role of pedigree selection in breeding of dairy cattle // Livestock breeding. № 2, 1985. - C. 19-23. Eisner F.F., Vlasov V.I., Podoba V.E. Genetic problems of selection of cattle // Vesnik s-kh nauki. - Л 3, 1965. - C. 73-78.

159. Ernst L.K. Increasing the efficiency of breeding - a powerful lever for the development of livestock // Dairy and beef cattle

breeding. - II, 1982. - c.19-23.

160. Ernst L.K. Study of the possibility of using genetic markers in animal breeding // Actual issues of applied genetics in animal breeding. - M.: Kolos, 1982. p. 92-94.

161. Ernst L.K., Sorokovoy P.F., Mashurov A.M. On the use of genetic markers tsri evaluation of genotypes of bulls-producers // Improvement of methods of breeding work with breeds of agricultural animals // Theses Dokl. conf. "Baisogola, 1973. - C.104-106.

162. Ernst L.K., Kabanov BD. New in the organization of pedigree business. - Moscow: Znanie, 1981. - C.33.

163. Ernst L.K., Tsalitis A.A. Large-scale selection in cattle breeding. - M.: Kolos, 1982. - C.238.

CONTENTS.

FSC
www.fsc.org
MIX
Papier | Fördert gute Waldnutzung
FSC® C083411

Kolibri 360 GmbH
Ferdinand-Jühlke-Straße 7
99095 Erfurt, Deutschland
produktsicherheit@kolibri360.de

Druck:
CPI Druckdienstleistungen GmbH
im Auftrag der
Kolibri 360 GmbH
Ferdinand-Jühlke-Str. 7
99095 Erfurt